PHYSIOLOGICAL CHEMISTRY

A Series Prepared under the General Editorship of

Edward J. Masoro, Ph.D.

I PHYSIOLOGICAL CHEMISTRY OF LIPIDS IN MAMMALS

II PHYSIOLOGICAL CHEMISTRY OF PROTEINS AND NUCLEIC ACIDS IN MAMMALS

III ENERGY TRANSFORMATIONS IN MAMMALS: Regulatory Mechanisms

IV ACID-BASE REGULATION: Its Physiology and Pathophysiology

In preparation

PHYSIOLOGICAL CHEMISTRY OF CARBOHYDRATES IN MAMMALS

REGULATION OF AMINO-ACID METABOLISM

ACID-BASE REGULATION: Its Physiology and Pathophysiology

Edward J. Masoro, Ph.D.

*Professor and Chairman, Department
of Physiology and Biophysics,
The Medical College of Pennsylvania*

Paul D. Siegel, M.D.

*Clinical Associate Professor of Medicine
The Medical College of Pennsylvania;
Director of Clinical Pulmonary Function and
Blood Gas Laboratory, Hospital of the
Medical College of Pennsylvania*

W. B. Saunders Company

Philadelphia · London · Toronto

W. B. Saunders Company: West Washington Square
Philadelphia, Pa. 19105

12 Dyott Street
London, WC1A 1DB

833 Oxford Street
Toronto, Ontario M8Z 5T9, Canada

Listed here is the latest translated edition of this book together with the language of the translation and the publisher.

Spanish (1st Edition) — Editorial Medica Panamericana, Buenos Aires, Argentina.

Acid-Base Regulation: Its Physiology and Pathophysiology   ISBN 0-7216-6143-2

Print No: 9   8   7   6   5   4   3

EDITOR'S FOREWORD

The past three decades or so have seen biochemistry emerge as possibly the most vigorous of the biological sciences. This, in turn, has led to a level of autonomy that has cut the cord linking biochemistry with its historically most important parent, mammalian physiology. For investigators in the fields of both biochemistry and physiology, this vitality has been most useful. But because of the arbitrary separation of these two disciplines in most teaching programs and all textbooks, the vast majority of students do not see the intimate relationships between them. Consequently, the medical student and the beginning graduate student as well as the recently trained physician find it difficult, if not impossible, to utilize the principles of biochemistry as they apply to the physiological and pathological events they observe in man and other mammals.

Therefore, this series is designed not only to introduce the student to the fundamentals of biochemistry but also to show the student how these biochemical principles apply to various areas of mammalian physiology and pathology. It will consist of six monographs: (1) Physiological Chemistry of Lipids in Mammals; (2) Physiological Chemistry of Proteins and Nucleic Acids in Mammals; (3) Energy Transformations in Mammals; (4) Acid-Base Regulation: Its Physiology and Pathophysiology; (5) Physiological Chemistry of Carbohydrates in Mammals; and (6) Regulation of Amino-Acid Metabolism.

The series can be profitably used by undergraduate medical students. Recent medical graduates and physicians involved in areas of medicine related to metabolism should find that the series enables them to understand the theoretical basis for many of the problems they face in their daily work. Finally, the series should provide students in all areas of mammalian biology with a source of information on the biochemistry of the mammal that is not otherwise currently available in textbook form.

EDWARD J. MASORO

PREFACE

In our experience in teaching acid-base regulation to medical students in both the preclinical and the clinical years, it has become evident that the major problem faced by the student is the correlation of the practical problems encountered in the clinic with the information learned in basic science courses. In our opinion this difficulty arises primarily from the fact that acid-base regulation is usually taught first from the viewpoint of the biochemist and then from the practical viewpoint of the clinician, with too little attention given to the physiology of acid-base regulation in normal man. We have, therefore, made the physiology of acid-base regulation in normal man the hub of this book and have related both biochemistry and clinical medicine to this basic physiologic approach. It is our belief that both the student who is first learning acid-base regulation and the physician who is involved in treating patients with acid-base disorders will find this approach of great value.

This book is organized in the following way: Chapter 1 is a broad overview of the entire subject; Chapters 2 through 4 provide the basic chemical and biochemical information underlying a consideration of acid-base physiology; Chapters 5 through 7 describe the physiological mechanisms involved in processing acids and bases in normal man; Chapters 8 through 11 discuss the pathophysiology of acid-base disturbances; and Chapter 12 presents a series of patients discussed on the basis of their pathophysiology relative to the material presented in the first eleven chapters. To our knowledge this is the first elementary text that starts with the basic chemistry of acid-base phenomena, continues through the physiology, and ends with a clinical view.

At the end of the first seven chapters, problems are presented. These problems should be solved by those students who feel that problem solving will enable them to better understand the material in the text. However, the text is so designed that students not requiring this problem-solving experience can omit the problems without interfering with their ability to comprehend the development of ideas. Appendix 2 contains solutions for all of the problems. If the student intends to solve the problems as part of his learning process, it is suggested that he first attempt to work a problem before referring to the appendix for the detailed method of solution.

The clinical approach to acid-base problems presented in the last five chapters is by no means the only approach to clinical disturbances of acid-base balance. We feel, however, that it is the most physiologically oriented approach and is to be preferred to a more chemically oriented approach when management of acid-base disorders is involved.

The references cited at the end of each chapter are not presented as documentation of the text but rather as a guide for the student who may wish to pursue selected areas of study more deeply. The works so listed are not necessarily the most important in the field but rather were selected because they present a broad discussion of a given area.

We wish to thank Miss Olivia Capers, Dr. Paul Kovnat, Dr. Elizabeth Labovitz, Miss Joan McGlinn, and Dr. William Sanslone for reading the text and providing us with criticisms from the point of view of both the student and the faculty member. We are most grateful to Mrs. Maureen Passe whose secretarial help made it possible for us to prepare the manuscript and to Mr. William Shriver whose help in preparing the illustrations was invaluable. Finally we are indebted to the editorial and art staff of the W. B. Saunders Company for their patience and skill in preparing the printed text.

E. J. MASORO
P. D. SIEGEL

CONTENTS

CHAPTER 9

METABOLIC ALKALOSIS ....

CHAPTER 10

CHAPTER 11

CHAPTER 12

1

INTRODUCTION

BODY FLUIDS

Acid-base regulation in mammals refers to those chemical and physiological processes which maintain the hydrogen ion (H^+) concentration in body fluids at levels compatible with life and proper functioning, i.e., good health. This is a sizable task because reactions which generate H^+ and reactions which consume H^+ are continuously occurring in mammals. To discuss the processes of acid-base regulation meaningfully, it is first necessary to consider body fluids broadly.

Water comprises about 60 per cent of the weight of the average human being, this percentage remaining rather constant in most individuals from day to day and month to month. Although constant for a given individual, the percentage of body weight due to water can vary markedly from individual to individual, ranging from 45 to 75 per cent. This broad range stems from the fact that the amount of adipose tissue relative to body weight differs greatly from person to person. Although water accounts for about 75 per cent of the weight of most tissues, it accounts for less than 10 per cent of the weight of adipose tissue. Obviously, therefore, the percentage of water content of a lean man is much higher than that of an obese man. A wide variety of organic and inorganic substances are dissolved in the body water, so that the term body fluids refers to the solvent water along with the solutes dissolved in it.

On an anatomical basis, body water is considered to be divided into two major compartments: the extracellular fluid compartment and the intracellular fluid compartment. These two compartments are separated by the plasma membrane or limiting membrane of the approximately 10^{13} cells of the human body. The fluid within the cells, termed the intracellular fluid, accounts for about 55 per cent of the body fluid and the fluid on the outside of the plasma membrane, termed the extracellular fluid, accounts for the remainder. Neither the extracellular fluid nor the intracellular fluid is a homogeneous solution; both contain subcompartments which vary in composition (Fig. 1-1).

The major subcompartments of the extracellular fluid are the blood plasma and the interstitial fluid. Blood plasma comprises about 4.5 per cent of the body weight, and is that portion of the extracellular fluid found within the confines of the blood vessels (i.e., the intravascular compartment of the extracellular fluid). The interstitial fluid comprises about 16 per cent of the body weight and is the subcompartment of the extracellular fluid outside of the vascular system which directly bathes the cells and is present in the lymphatic system, the latter fluid being specifically called lymph. The plasma and interstitial fluids are separated from each other by the endothelium of the blood capillaries. These endothelial cells form a membranous structure which permits substances of low molecular weight to pass across it but markedly restricts the passage of macromolecules. Although there is a slow drainage of lymph into the venous system by way of ducts, the main interactions between interstitial fluid and blood plasma occur at the endothelial membrane of the blood capillaries.

The composition of plasma and interstitial fluid are presented in Table 1-1. Plasma contains much protein, while the interstitial fluid contains little. This difference results from the fact that the endothelium of the blood capillaries rather effectively restricts the passage of proteins from plasma to interstitial fluid. However, the endothelial membranes are not perfect ultrafilters, thus the value of zero for the protein content of interstitial fluid recorded in Table 1-1 is not strictly true. Small amounts of protein pass across even tight capillary membranes such as those found in skeletal muscle, and in a tissue such as liver where blood capillaries have a sinusoidal structure, the interstitial fluid contains a considerable amount of protein.

Plasma proteins are polyanions which means that the endothelium separates a fluid containing a fairly high concentration of a nondiffusible

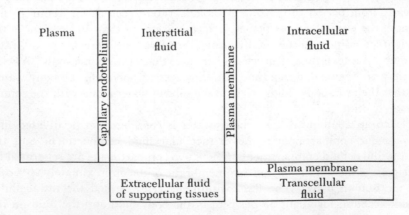

Figure 1-1. Schematic representation of body fluid compartments.

Table 1-1. *Concentrations of Cations and Anions In Plasma Water and Interstitial Fluid*

	ION	PLASMA* (mEq. per liter)	PLASMA WATER (mEq. per liter)	INTERSTITIAL FLUID† (mEq. per liter)
CATIONS	Na+	142	152.7	145.1
	K+	4.0	4.3	4.1
	Ca++	5.0	5.4	3.5
	Mg++	2.0	2.2	1.3
	Total	153.0	164.6	154.0
ANIONS	Cl−	102.0	109.9	115.7
	HCO$_3^-$	26.0	27.9	29.3
	PO$_4^{\equiv}$	2.0	2.1	2.3
	Other	6.0	6.5	6.7
	Protein	17.0	18.2	0.0
	Total	153.0	164.6	154.0
Total mOsm. per liter		306.0	329.2	308.0

* A plasma water content of 93 per cent was used in the calculation.

† Gibbs-Donnan factors used are 0.95 for monovalent anions and 1.05 for monovalent cations.

(From Woodbury, D. M.: *In* Ruch, T. C., and Patton, H. D. (eds.): Physiology and Biophysics. W. B. Saunders Co., Philadelphia, 1965.)

anion from a fluid containing a low concentration of such anions. Therefore it is to be expected that the interaction between plasma and interstitial fluid might approach a Gibbs-Donnan equilibrium.* (Of course normally a steady-state is actually maintained but one which is markedly influenced by the Gibbs-Donnan effect.)

It will be noted (Table 1-1) that the concentration of monovalent cations† in plasma water is higher than in interstitial fluid, and the concentration of diffusible anions in the plasma water is lower than in the interstitial fluid. The extent of these differences are in accord with the

---

* The Gibbs-Donnan equilibrium occurs in systems containing two fluid compartments separated by a semipermeable membrane which permits free passage of some ions in solution and prevents passage of others. If the concentration of nondiffusible ions in each compartment differs, then at equilibrium the two fluid compartments will differ in regard to the distribution of the diffusable ions. There will also be osmotic and electrical differences between the two fluid compartments. Indeed, this is a complex phenomenon, and those interested in it can review it in a physical chemistry textbook (one of which is listed in the references at the end of this chapter). From the point of view of our discussion of acid-base regulation, it is only necessary for the reader to know that plasma proteins are relatively nondiffusible polyanions relative to the capillary endothelial membrane. Therefore, because of the Gibbs-Donnan effect a distribution of the diffusible ions such as Cl−, HCO$_3^-$, K+, and Na+ occurs, resulting in a higher concentration of diffusible cations in the plasma than in the interstitial fluid and just the opposite for the diffusible anions.

† In Tables 1-1 and 1-2, the concentrations are expressed in mEq. per liter. In the case of monovalent ions the number of milliequivalents of an ion per liter is equal to the number of millimoles per liter, but for divalent ions each milliequivalent of the ion is equal to 0.5 millimoles of the ion.

Table 1-2.   *Intracellular Concentrations of Cations and Anions in Some*
Representative Tissues

ION		SKELETAL MUSCLE	CARDIAC MUSCLE	LIVER	THYROID GLAND	ERYTHROCYTES
		(mEq. per liter intracellular water)				
CATIONS	Na^+	12	7	3	42	19
	K^+	150	134	148	147	136
	Mg^{++}	34	28	31	?	6
	Ca^{++}	4	4	2	?	0
	Total	200	173	184		161
ANIONS	Cl^-	4	4	16	19	78
	HCO_3^-	12	12	?	14	18
	$PO_4^{\equiv+}$	40				4
	Organic					
	Protein	54				36
	Total	110				136
Total mOsm. per liter		310				310
E.C.F. volume		8.2 per cent	19.5 per cent	15 per cent	30 per cent	—

(From Woodbury, D. M.: *In* Ruch, T. C., and Patton, H. D. (eds.): Physiology and
Biophysics. W. B. Saunders Co., Philadelphia, 1965.)

Gibbs-Donnan effect playing a major role. However, the higher concentration of the divalent cations Ca^{++} and Mg^{++} in plasma relative to interstitial fluid does not relate primarily to the Gibbs-Donnan effect but rather
to the binding of these cations by the plasma proteins.

There is a rapid exchange of water and diffusible solutes between
plasma and interstitial fluid. However, there is another subcompartment
of the extracellular fluid, found within the supporting tissues (such as con-
nective tissue and bone), which does not show rapid exchange of these
substances with those of the major subcompartments; the exact size and
dynamics of this subcompartment have yet to be worked out.

Finally, a small volume of the extracellular fluid, comprising 1 to 3 per
cent of the body weight, is termed the transcellular fluid. This refers to
fluids formed by the secretory activity of specific epithelial cells, e.g., the
cerebrospinal fluid, fluids secreted into the lumen of the gastrointestinal
tract, the intraocular fluids, and others. The composition of these fluids can
markedly differ from that of the bulk of the extracellular fluid, e.g., the
parietal cells of the gastric glands secrete a fluid containing about 0.1 N HCl.

The intracellular fluid is often spoken of in such a way that one might
feel it is a homogeneous fluid; nothing could be further from the truth. In
Table 1-2 the composition of the intracellular fluid of several tissues is

compared, and a glance reveals that the intracellular fluid of skeletal muscle, for example, is different in composition from that of other tissues. Not only does such a difference exist between different tissues, but there are sub-compartments of vastly different compositions within the same tissue, or for that matter, the same cell. For instance, in the resting (noncontracting) skeletal muscle cell, most of the Ca^{++} is in the sarcoplasmic reticulum membrane system with almost none in the fluid of the sarcoplasm surrounding the myofibrils. Obviously intracellular fluid is most heterogeneous, with the fluid of each kind of cell differing from that of other kinds of cells and with many fluid subcompartments existing within the same cell. This heterogeneity must be recognized when considering acid-base regulation; yet it is often useful and sufficiently correct for the purposes at hand to consider the intracellular fluid as a single fluid compartment.

If the intracellular and extracellular fluids are considered as homo-geneous fluids, the generalization can be made that the major cations quantitatively of intracellular fluid are K^+ and Mg^{++} and that of the extracellular fluid is Na^+. The major anions of the intracellular fluid are proteins and organic phosphates and those of the extracellular fluid Cl^- and HCO_3^- (see Tables 1-1 and 1-2). These differences between intra-cellular and extracellular fluids involve selective permeability of the plasma membrane and transport systems (or pumps) located in the plasma membranes of the cells.

It is believed that water can pass freely across most biological mem-branes. Therefore it is rather generally accepted that the osmolarity (osmotic pressure) of all of the various fluid compartments is similar. There are some exceptions to this generalization, a most important one being the somewhat higher osmotic pressure of plasma compared to other body fluids, but by and large it appears to be a reasonable supposition. There is one region with a markedly higher osmolarity than that of most body fluids, namely, the medulla of the kidney, but this is a special case which has little direct importance in the consideration of acid-base regulation.

H^+ CONCENTRATION OF BODY FLUIDS

Besides the cations listed in Table 1-1 and Table 1-2, body fluids contain hydrogen ion, H^+. Taken literally H^+ refers to a hydrogen atom without its orbital electron, i.e., H^+ refers to a proton. A proton is a most unique ion since it has an effective radius of about 10^{-13} cm compared to that of approximately 10^{-8} cm for most other simple ions, e.g., Na^+, Cl^-, and so forth. However, little or no H^+ in the form of naked protons exists in aqueous solutions because of the enormously high ratio of charge to the radius of such an ion. Almost all protons (H^+) in aqueous solutions are reacted with H_2O to form hydrated ions such as H_3O^+, often called hydro-nium ions. Actually in the body fluids, most of the hydronium ions in solution are probably in the form of $(H_5O_2)^+$, $(H_7O_3)^+$, $(H_9O_4)^+$, and ions of even greater hydration; the extent of hydration depends upon factors such as the concentration of ions and the temperature. Throughout this text the symbol H^+ and the term hydrogen ion will be used as a convenience, but the reader should realize that what is meant by the concentration of H^+ in body fluids is hydrated protons or hydronium ions.

The fluid compartment of the body most studied in regard to H^+ concentration is the blood plasma. In normal people the concentration of H^+ is approximately 40 nanomoles (nmoles)* per liter of plasma. Compared to other cations found in plasma, therefore, the concentration of H^+ is very small indeed. For instance, the plasma contains per liter approximately 142 million nmoles of Na^+, 4 million nmoles of K^+, 2.5 million nmoles of Ca^{++}, and 1 million nmoles of Mg^{++}. Clearly the H^+ concentration of plasma is about one-millionth plus or minus one order of magnitude of that of the major monovalent and divalent cations of plasma. A similar quantitative relationship exists between H^+ and the major anions in plasma, i.e., Cl^-, HCO_3^-, phosphate, and protein. That the H^+ concentration of plasma and other body fluids is about one-millionth that of other ions is a most important fact to bear in mind when considering acid-base regulation.

The H^+ concentration of the interstitial fluid is slightly less (about 5 per cent less) than that of the plasma. This is true because H^+, like other diffusible cations, is distributed between plasma and interstitial fluid in accordance with the Gibbs-Donnan membrane effect.

Precise information on the H^+ concentration of the intracellular fluid is difficult to find. No doubt each tissue type has a different H^+ concentration than other tissue types. Moreover, because of the heterogeneity of the intracellular fluid within a given cell, it is likely that the H^+ concentration will be different in each subcellular compartment. For instance, it is to be expected that the H^+ concentration at membrane surfaces is different from that in the so-called particle-free fluid or cytosol of the cell. Nevertheless it is useful to make an approximate, average estimate of H^+ concentration of intracellular fluid; about 100 nmoles of H^+ per liter of intracellular fluid would seem to be the figure most agreed on, but some feel that 1000 nmoles per liter is closer to the truth. The methods used to measure the intracellular H^+ concentration, the reasons for disagreements about its concentration, and the probable forces involved in the regulation of its concentration will be discussed in Chapter 3.

CONCEPT OF pH

In 1909 Sørensen developed the pH scale for expressing H^+ concentrations ranging from 1 M to 10^{-14} M. He defined the pH of a solution

---

* 1 nmole equals 10^{-9} moles.

as the negative value of the logarithm of its H^+ concentration, algebraically expressed as follows:

$$pH = -\log [H^+]$$

The symbol $[H^+]$ refers to H^+ concentration in moles per liter or equivalents per liter. The pH scale, as commonly used, ranges from 0 for a 1 M H^+ solution to 14 for a 10^{-14} M solution.

As the concept of the activity of ions in solution gained wide acceptance and understanding, the definition of pH was reinterpreted as follows:

$$pH = -\log \alpha_H$$

The symbol α_H refers to the activity of the H^+.

In recent years it has been recognized that pH is not an absolute measure of H^+ concentration or activity but rather a measure relative to a standard. (For further discussion of the meaning of pH see reference to Bates cited at the end of this chapter.) Because of this, some investigators feel it to be totally incorrect to translate pH into either concentration or activity terms. However, from a practical biomedical viewpoint it is useful to define pH as follows:

$$pH \cong -\log [H^+]$$

Provided it is realized that only an approximate idea of the absolute value of the H^+ concentration is gained from pH measurements, translation of pH into $[H^+]$ terms aids in the discussion of acid-base regulation.

Expression of the H^+ concentration in terms of pH has certain advantages. For one thing, it is a convenient way of expressing a wide range of concentrations. For example, for the following solutions, 0.1 N HCl, pure H_2O, blood plasma, and a 0.1 N NaOH solution, it is simpler to express their H^+ concentrations as pH 1, 7, 7.4, and 13 respectively than as molar concentrations of 0.1 M, 10^{-7} M, 4×10^{-8} M, and 10^{-13} M respectively. Furthermore, there are certain advantages to the pH scale (which will become evident in Chapter 2) when considering buffers.

Nevertheless, for physicians and physiologists the pH system also presents certain disadvantages. One of these is the inability of most people to think in a facile way when using a system that both inverts and uses a log scale to express H^+ concentration. For instance it is not immediately apparent that a change in pH from 7.7 to 7.4 represents a doubling of the H^+ concentration. It is also not immediately obvious that a plasma pH of 7.4 means that the H^+ concentration of plasma is less than 1 millionth that of Na^+. Since the range of plasma H^+ concentrations compatible with life is within the nanomolar (nM) concentration range, the H^+ concentration throughout much of this text will be expressed in terms of nmolarity or nmoles/L. However, pH will be used too, particularly in those areas where it will be easier to understand the chemical system. Also, since pH is widely used by chemists and physicians, in many instances H^+ concentration will be expressed both in terms of nmolarity and of pH so that the reader will gain a familiarity with both systems.

SURVEY OF THE REGULATION OF H⁺ CONCENTRATION
OF THE BODY FLUIDS

The H^+ concentration of the body fluids of the normal individual does not vary markedly even though many processes are occurring that might be expected to change it greatly. For instance, foodstuffs such as glucose and triglyceride which in the body fluids do not behave as acids, in that they do not donate H^+ to these fluids, are converted to CO_2 during the course of their metabolism, and much of this CO_2 is hydrated to carbonic acid (H_2CO_3) which dissociates to yield H^+ and HCO_3^-.

About 13,000 millimoles (mmoles) of CO_2 are formed each day but the body processes all of this carbonic acid without changing the H^+ concentrations of the body fluids appreciably. This is accomplished in the following way: First, most of the H^+ from CO_2 does not remain as such in solution but is immediately removed by the action of chemical buffers (see Chapter 2). Secondly, the ventilatory component of the respiratory system eliminates CO_2 from the body at the same rate at which it is formed, thus maintaining a rather constant steady-state concentration of CO_2 in the body fluids. In summary there are two important systems maintaining the relative constancy of the $[H^+]$ in the face of massive CO_2 production: (1) chemical buffering and (2) elimination of CO_2 by alveolar ventilation.

The ventilatory system by eliminating CO_2 is not only of importance in maintaining an appropriate relationship between the production and excretion of CO_2 but it also functions in a compensatory fashion when acids (other than carbonic acid) and alkalis threaten to markedly alter the H^+ concentration of the body.

The metabolism of foodstuffs can yield other acids besides carbonic acid, e.g., lactic acid from glucose or glycogen, acetoacetic acid and β-hydroxybutyric acid from triglyceride, phosphoric acid from phospholipids, and sulfuric acid from proteins, all of which dissociate to yield significant amounts of H^+. Much of this acid merely causes transient acid-base problems because the further metabolism of lactic, β-hydroxybutyric, and acetoacetic acids converts them to CO_2 (which is readily excreted as previously described) and H_2O. Moreover, during the transient period of accumulation of such acids, the H^+ which they yield is buffered by the chemical buffers of the body fluids. If the accumulation of these acids causes the H^+ concentration to increase significantly an increased rate of elimination of CO_2 by alveolar ventilation tends to restore the H^+ concentration toward a normal level, i.e., respiratory compensation takes place. Moreover the kidney will slowly excrete the excess H^+.

Phosphoric and sulfuric acids cannot be further metabolized to CO_2. Therefore the elimination of these acids and the H^+ which they yield can only be accomplished by the kidneys which restore, simultaneously with the elimination of H^+, the buffer systems of the body that had been used to buffer the H^+ released from these acids. The kidney of a normal man excretes about 50 milliequivalents of H^+ per day in urine.

The metabolism of foodstuffs does not always tend to acidify the body fluids; some foodstuffs have an alkalizing action. For instance, the ingestion of large amounts of salts of organic acids (i.e., Na^+ or $K^+ + R - COO^-$) found in fruit (e.g., sodium lactate, sodium citrate, and so forth) alkalinize the body fluids because the organic anions are metabolized to CO_2 and water, a process that consumes one mole of H^+ for every mole of organic anion metabolized. The buffer systems and respiratory compensation prevent a marked fall in H^+ concentration, and ultimately the kidney excretes the alkali and in the process readjusts the buffer systems of the body fluids.

In summary, the body is capable of maintaining a fairly constant H^+ concentration in its fluids by means of three mechanisms: (1) the chemical buffer systems of the body fluids, (2) the capacity of alveolar ventilation to eliminate CO_2 as quickly as it is formed, plus its ability to alter the rate of CO_2 elimination in a compensatory fashion relative to H^+ concentration of the body fluids, and (3) the capacity of the kidneys to eliminate acid or alkali and in so doing readjust the buffer systems of the body. The primary purpose of this book is to explore each of these mechanisms in depth and to integrate them in terms of the physiology of man and other mammals and then to discuss the system under pathophysiologic conditions.

BIOLOGICAL IMPORTANCE OF REGULATION OF H^+ CONCENTRATION

H^+ (i.e., hydronium ions) are much more reactive than the other cations in body fluids because the small radius of H^+ permits very strong interactions between H^+ and negatively charged regions of other molecules. Such interactions are most important when macromolecules such as proteins are being considered because the interaction of H^+ with negatively charged functional groups of proteins leads to changes in the charge distribution of the proteins and therefore to marked changes in conformation.

Since both the conformation of a protein enzyme and the charge on its functional groups greatly influence its enzymatic activity, it is to be expected that changes in H^+ concentration should markedly affect the catalytic activity of enzymes. Indeed such is the case; the activity of most enzymes becomes maximal at a well defined H^+ concentration (the so-called pH optimum of the enzyme) and the enzyme is inactive when the H^+ concentration of the environment is markedly different than this optimum. Other physiologic activities of macromolecules (e.g., blood clotting, muscle contraction) are similarly influenced by the H^+ concentration as expected since such activities are dependent on conformation of the macromolecule and the charge of certain of its functional groups.

Actually it is not necessary to theorize about the biological importance of careful regulation of H^+ concentration on the basis of macromolecular chemistry and cell physiology to gain a full appreciation of its significance to the organism, because both clinical and experimental evidence shows that the plasma H^+ concentrations of a mammal must be in the range of about 20 nmoles to 160 nmoles/L for life to continue (the normal value being 40 nmoles/L) and that at the extremes of this range the animal is very near death. It is true that on a percentage basis the mammal can tolerate (but with difficulty) rather large alterations in the hydrogen ion concentration of the plasma. However it should be remembered that even the upper limit of 160 nmoles/L is a negligible concentration of cation compared to the concentration of the other important ions in plasma such as K^+, Na^+, Cl^-, and HCO_3^-. Normally the plasma H^+ concentration is maintained fairly close to 40 nmoles/L and such regulation of H^+ concentration is necessary for the proper functioning of the organism. Although direct evidence is not as available as for plasma, it is rather certain that a similar regulation of the intracellular $[H^+]$ is also required for survival.

PROBLEMS

1. If the pH of a solution containing $[H^+]$ of 10^{-5} M is 5, what is the pH of a solution with twice that $[H^+]$?

2. What is the H^+ concentration expressed in nmoles per liter of a pH 7.6 solution?

3. What is the pH of a solution containing 150 mEq of H^+ per liter?

REFERENCES

For Discussion of Body Fluids:

Woodbury, D. M.: Physiology of body fluids. *In* Ruch, T. C., and Patton, H. D. (eds.): Physiology and Biophysics. W. B. Saunders Co., Philadelphia, 1965, pp. 871–898.

For Discussion of Gibbs-Donnan Membrane Equilibrium:

Williams, V. R., and Williams, H. B.: Basic Physical Chemistry for the Life Sciences. W. H. Freeman and Co., San Francisco, 1967, pp. 100–103.

For Discussion of H^+ *and Hydronium Ions:*

Vanderwerf, C. A.: Acids, Bases and the Chemistry of the Covalent Bond. Reinhold Publishing Corp., New York, 1961.

For Discussion of the Intracellular H^+ *Concentration:*

Bittar, E. E.: Cell pH. Butterworth Inc., Washington, D.C., 1964.
Butler, T. C., Waddel, W. J., and Poole, D. T.: The pH of intracellular water. Ann. N. Y. Acad. Sci. *133*:73–77 (1966).

Siesjö, B. K., and Pontén, U.: Intracellular pH—true parameter or misnomer. Ann. N.Y. Acad. Sci. *133*:78–86 (1966).

Waddell, W. J., and Bates, R. G.: Intracellular pH. Physiol. Rev. *49*:285–329 (1969).

For Discussion of pH:

Bates, R. G.: Determination of pH: Theory and Practice. John Wiley & Sons, New York, 1964.

Clark, W. N.: Topics in Physical Chemistry. Williams and Wilkins Co., Baltimore, 1952.

For Discussion of Effect of H+ *on Macromolecules:*

Kaldor, G.: Physiological Chemistry of Proteins and Nucleic Acids in Mammals. W. B. Saunders Co., Philadelphia, 1969.

2

CHEMISTRY OF ACIDS, BASES, AND BUFFERS

THE MEANING OF THE TERMS ACID AND BASE

Over the centuries there have been many concepts regarding the meaning of the terms acids and bases. Indeed, there are today several ways of defining acids and bases. For physiological and medical purposes, the Brønsted-Lowry concept of acids and bases is the most useful and will be the one used throughout this text. However, because other concepts of acids and bases, even archaic ones, appear in the literature, a brief review of some of these concepts is presented.

The ancients described acids as sour-tasting substances and defined bases as substances that "killed" acids. Since the carbonates and hydroxides of alkali metals were the first bases in wide use, the term alkali came to be used synonymously with the term base. Although some modern chemists prefer to restrict the use of the term alkali to bases containing alkali metals, base and alkali will be used as synonyms in this text.

In 1887, Arrhenius proposed that the characteristics of acids in water solutions were due to the properties of hydrogen ions (H^+) and those of bases to the properties of hydroxide ions (OH^-). By this definition an acid is a substance whose water solution contains more H^+ than OH^-, and a base is a substance whose water solution contains more OH^- than H^+. Although this concept was an important milestone, it is too restricted in scope to serve as the basis for a discussion of acid-base regulation. *

---

* However, it should be noted that by using a system for CO_2 chemistry other than the one in general use, MacConnachie recently presented a comprehensive view of acid-base regulation based on the Arrhenius concept. This approach has no real advantage over the more widely used Brønsted-Lowry concept; moreover, until completely mastered, a single approach should be used by the student. For those who are well acquainted with acid-base regulation, however, it is well worth reading the article by MacConnachie (see references).

Over the years there have been several other ways of defining a base, one of which, although poor, should be mentioned because it is sufficiently used by physicians to warrant discussion. It probably relates to the long use of the term base in describing the principal ingredient of a compound. This led to naming the metal components of salts as the bases or principal components. By this definition, Na^+ and K^+ are called bases. However, to call these cations bases in the context of acid-base physiology does not lead to a logical development of the subject. Therefore, although this definition of base must be understood since it will be encountered, its use should be discouraged.

Ideally a single system of defining acids and bases should be used by all workers in the biomedical field, and the one chosen should best suit the needs of this specialized area of study. The Brønsted-Lowry concept of acids and bases, properly used, fits this bill nicely and for this reason has been widely adopted. In this system, an acid is defined as a substance that can donate a proton to another substance. A base is defined as a substance that can accept a proton from another substance. In short, an acid is a proton donor and a base is a proton acceptor.

Before turning to exclusive use of the Brønsted-Lowry concept, brief mention should be made of the Lewis concept of acids and bases, since the modern organic chemist finds it most useful. Lewis defined an acid as an electron-pair acceptor and a base as an electron-pair donor. In this system an acid-base reaction is a base sharing an electron pair with an acid. This system provides the organic chemist with an important tool for understanding organic reactions and for considering acid-base reactions in non-aqueous solvents. However, for the physiologist and physician dealing with animal systems, the very breadth of the Lewis concept is a disadvantage. Indeed for physiological systems, it is necessary to restrict the domain of the Brønsted-Lowry system (as discussed in Chapter 3) to gain the best understanding of biomedical acid-base problems.

STRONG ACIDS, WEAK ACIDS, AND BASES

Since the body is an aqueous system, our discussion of acids and bases will be restricted to conditions where H_2O is the solvent. In aqueous systems acids can be classified into two groups: (1) strong acids and (2) weak acids. There is no sharp boundary between the two groups; rather there are all gradations between an obviously strong acid, such as hydrochloric acid, and a clear-cut weak acid, such as acetic acid. A strong acid is defined as one which is completely or almost completely ionized in water solution. A weak acid is defined as one which is only slightly ionized in aqueous solution. The terms strong and weak should not be confused with acid concentration, e.g., a 10^{-5} M HCl solution is a dilute solution of a strong acid and a 5 M acetic acid solution is a concentrated solution of a weak acid.

Hydrochloric, perchloric, nitric, and sulfuric acids are examples of commonly encountered strong acids. The interaction of a strong acid with an aqueous environment is symbolized in Equation 2-1 in which hydrochloric acid is the example:

$$HCl + H_2O \rightarrow H_3O^+ + Cl^- \qquad (2\text{-}1)$$

The arrow in Equation 2-1 points to the right; there is no arrow pointing left. This is done to emphasize that there is almost total ionization of hydrochloric acid in an aqueous environment, but it is not strictly true since probably a minute amount of undissociated HCl is present in most aqueous solutions containing H_3O^+ and Cl^-. Hydrochloric acid clearly fits the definition of Brønsted-Lowry acid since it donates its protons to water to form hydronium ions. It is also apparent from Equation 2-1 that water in this case is functioning as a Brønsted-Lowry base, i.e., it is a proton acceptor.

Equation 2-1 provides a total picture of the reaction of HCl with H_2O. If the shorthand notation of H^+ is used for hydronium ion (as discussed in Chapter 1) then the reaction is written as in Equation 2-2:

$$HCl \rightarrow H^+ + Cl^- \qquad (2\text{-}2)$$

Although Equation 2-2 is an excellent and widely used shorthand, it must always be realized that in aqueous systems acids donate their protons to the solvent water. Therefore when encountering equations like Equation 2-2, it is important to envision the full picture as shown by Equation 2-1.

During the course of this text many different weak acids will be discussed. Acetic acid is a good example of such acids, and its behavior in water solution is symbolized by Equation 2-3:

$$CH_3COOH + H_2O \underset{\longleftarrow}{\overset{\rightarrow}{\rule{0pt}{0pt}}} H_3O^+ + CH_3COO^- \qquad (2\text{-}3)$$

Acetic acid donates protons to water to yield hydronium and acetate ions. In Equation 2-3 the arrow from left to right is drawn quite short and the one from right to left quite long in order to indicate that at equilibrium the system contains far more undissociated acetic acid than acetate or hydronium ions. Equation 2-3 can be expressed in short-hand by Equation 2-4:

$$CH_3COOH \underset{\longleftarrow}{\overset{\rightarrow}{\rule{0pt}{0pt}}} H^+ + CH_3{-}COO^- \qquad (2\text{-}4)$$

For a general discussion of weak acids, the symbol HA will be used to denote the undissociated acids. Their reaction with water is depicted in Equation 2-5:

$$HA + H_2O \underset{\longleftarrow}{\overset{\rightarrow}{\rule{0pt}{0pt}}} H_3O^+ + A^- \qquad (2\text{-}5)$$

This equation shows that the reaction going from left to right involves the weak Brønsted-Lowry acid, HA, and the Brønsted-Lowry base, H_2O. The reaction proceeding from right to left involves the Brønsted-Lowry acid, hydronium ion, and the Brønsted-Lowry base, A^-. By use of the Law of Mass Action it is possible to calculate the velocity of the reaction proceeding from left to right and the one from right to left. The velocity from left to right, denoted by the symbol v_1, is related to the concentration of reactants by Equation 2-6:

$$v_1 = k_1[HA][H_2O] \qquad (2\text{-}6)$$

In other words, v_1 is equal to the product of the molar concentration of the two reactants multiplied by k_1, the rate constant* for this particular reaction. The velocity of the reaction from right to left, denoted by v_2, is expressed in a similar manner by Equation 2-7:

$$v_2 = k_2[H_3O^+][A^-] \qquad (2\text{-}7)$$

Again in words, v_2 equals the product of the molar concentrations of the reactants multiplied by k_2, the rate constant for the reaction proceeding from right to left in Equation 2-5.

If 1 mmole of HA is added to 1 liter of water the initial velocity of the reaction from left to right can be calculated from Equation 2-6; as the reaction takes place the concentration of HA declines and thus v_1 will also decline. Moreover as A^- and H_3O^+ form, their concentrations increase, and therefore v_2, which initially was zero, increases. Finally a concentration of reactants on both sides of Equation 2-5 will be reached where v_1 and v_2 are equal; at this point the system is in equilibrium, i.e., there is no change in concentration of reactants with time. The equilibrium situation can be expressed algebraically by Equation 2-8:

$$v_1 = v_2 \qquad (2\text{-}8)$$

This equilibrium state and the equations that can be derived from it are important in acid-base chemistry and in the understanding of buffers. The following algebraic manipulations are carried out to yield useful expressions from Equation 2-8, i.e., from the equilibrium state. By inspecting Equations 2-6 and 2-7 it is obvious that Equation 2-8 can be reexpressed as shown in Equation 2-9:

$$k_1[HA][H_2O] = k_2[H_3O^+][A^-] \qquad (2\text{-}9)$$

Equation 2-9 can be algebraically rearranged to yield Equation 2-10:

$$\frac{k_1}{k_2} = \frac{[H_3O^+][A^-]}{[HA][H_2O]} \qquad (2\text{-}10)$$

The left hand term of Equation 2-10 is a ratio of the two constants, k_1 and k_2; the ratio of two constants can be set equal to a third constant designated K, called the equilibrium constant, as shown in Equation 2-11:

$$\frac{k_1}{k_2} = K \qquad (2\text{-}11)$$

---

* The rate constant is defined for a specified temperature; i.e., its numerical value changes as the temperature changes.

The left hand term in Equation 2-10 can be replaced by K of Equation 2-11 to yield Equation 2-12:

$$K = \frac{[H_3O^+][A^-]}{[HA][H_2O]} \qquad (2\text{-}12)$$

If the concentration of the reactants of Equation 2-5 are known under equilibrium conditions, K, the equilibrium constant of the weak acid, HA, can be calculated by Equation 2-12, but only for the theoretical condition of infinite dilution where interionic forces between solute molecules are negligible. However, an equation of the form of Equation 2-12 has such great usefulness that, if possible, its form should be maintained in an equation of practical use. Equation 2-13 meets this requirement:

$$K' = \frac{[H^+][A^-]}{[HA]} \qquad (2\text{-}13)$$

K′, called the apparent ionization or dissociation constant, encompasses the activity coefficient and eliminates the water term from the denominator of the right hand side of Equation 2-12. In Equation 2-13, H_3O^+ is replaced by the shorthand symbol H^+. K′, of course, is not really a constant but can be used as one in a system limited to specifically defined ionic strength and temperature. For any other ionic strength or temperature, a different numerical value for the K′ must be used. Since both the ionic strength and the temperature of body fluids are known and are maintained reasonably constant, Equation 2-13 is most valuable for considering acid-base physiology.

An immediately obvious significance of K′ is the information that it provides in regard to the strength of an acid. The higher the value of the K′, the stronger is the acid. For example, the K′ of hydrochloric acid is a very large number indeed, fitting the concept that hydrochloric acid is a strong acid. In contrast, the K′ of acetic acid is around 10^{-5}, the exact value depending on the ionic strength and temperature of the system. Much use will be made of Equation 2-13 and of the concept of the K′ when buffers are discussed later.

The reader may be wondering why there has been so much mention of H^+ and almost none of OH^-. Indeed this will be true throughout the book because once the $[H^+]$ of an aqueous solution is known, the $[OH^-]$ is easily calculated. To understand why this is true it is necessary to consider the dissociation of pure water which occurs to a limited extent as shown in Equation 2-14:

$$H_2O + H_2O \xrightarrow{\longleftarrow} H_3O^+ + OH^- \qquad (2\text{-}14)$$

In the reaction proceeding from right to left, two water molecules react to yield hydronium (or H^+) and OH^- ions. Water in this reaction is functioning both as an acid and a base in the Brønsted-Lowry sense; i.e., one molecule of water donates a proton to another molecule of water which accepts it. Like any other weak-acid system, the dissociation of water can be quantitatively considered by Equation 2-12 or more specifically by Equation 2-15:

$$K = \frac{[H_3O^+][OH^-]}{[H_2O]^2} \qquad (2\text{-}15)$$

Algebraic rearrangement yields Equation 2-16:

$$[H_3O^+][OH^-] = K[H_2O]^2 \qquad (2\text{-}16)$$

Since the concentration of H_2O in dilute solutions, such as body fluids, is fairly constant, it is proper to express the right hand side of Equation 2-16 as a constant (the multiple of two constants $[H_2O]^2$ and K) symbolized as K_w (often called the dissociation constant or the ionization constant or the ion product of water) as shown in Equation 2-17:

$$K[H_2O]^2 \cong K_w \qquad (2\text{-}17)$$

K_w can be used in place of the right hand term in Equation 2-16 to yield Equation 2-18:

$$[H_3O^+][OH^-] = K_w \qquad (2\text{-}18)$$

The numerical value of K_w varies with temperature and other properties of the solution but is approximately 10^{-14}. As calculated from Equation 2-18 pure H_2O contains 10^{-7} moles/L hydronium ion (or H^+) and 10^{-7} moles/L OH^-. Moreover Equation 2-18 enables the $[OH^-]$ to be rapidly calculated whenever $[H^+]$ is known. For example, a 0.1 M HCl solution contains approximately 10^{-1} moles H^+ per liter; from Equation 2-18, it can be calculated that the solution also contains 10^{-13} moles OH^- per liter.

TITRATION CURVES

Further understanding of the properties of strong and weak acids can be gained by considering the titration curves of typical examples of such acids as shown in Figure 2-1. The data presented relate to the titration of 1 liter of 0.1 M solutions of hydrochloric acid (as an example of a strong acid) and acetic and boric acids (as typical examples of weak acids). The titrant used is 10 M sodium hydroxide; a base of this concentration is used because a 10-ml volume contains the same number of millimoles of the base as the millimoles of acid in the 1 liter of a 0.1 M acid solution. Therefore the titration of these acids with this base does not cause a significant volume change in the system, thus circumventing certain complications in the theoretical analysis of the data. (Of course from a technical point of view titration with such a concentrated base would not yield data of highest accuracy.) In Figure 2-1 the mls of 10 M sodium hydroxide titrant added is plotted on the Y axis and the pH of the system on the X axis.

A 0.1 M solution of hydrochloric acid has a pH $\cong$ 1 since it is a strong (i.e., fully dissociated) acid. The extent to which the pH differs from 1

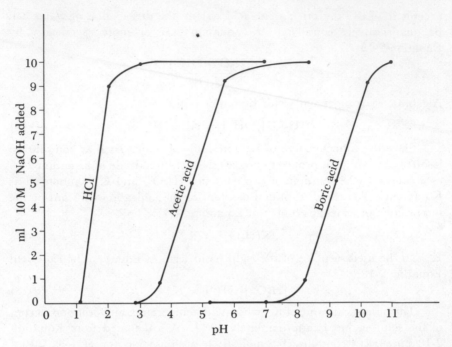

Figure 2-1. Titration curves for strong and weak acids. The data presented are for 1 liter of 0.1 M solutions of acid.

relates to the H^+ activity of a 0.1 M HCl solution; however, except for plotting the pH of this 0.1 M solution as being somewhat greater than 1, the effect of activity is ignored in this discussion. The addition of 9 ml of the titrant to the liter of 0.1 M HCl involves reaction of OH^- with 90 per cent of the H^+ (as follows: $H^+ + OH^- \rightarrow H_2O$); however, since 10 per cent of the H^+ still remains, a $pH \cong 2$ is reached (see Fig. 2-1). The addition of 9.9 ml of titrant involves reaction of OH^- with approximately 99 per cent of the H^+ of the HCl solution; since only 1 per cent of the H^+ remains, a $pH \cong 3$ is reached (see Fig. 2-1). The addition of 10 ml of titrant involves reaction of all the H^+ of the HCl with OH^-, resulting in NaCl in aqueous solution ($pH \cong 7$). The curve drawn in Figure 2-1 for the data just discussed is a titration curve typical of any strong acid.

It will be noted in Figure 2-1 that a 0.1 M acetic acid solution exhibits a considerably higher pH than the 0.1 M HCl solution. The difference in pH relates primarily to the fact that HCl is a strong or fully dissociated acid, while acetic acid is a weak or partially dissociated acid. Addition of titrant rapidly consumes the small amount of H^+ in solution. Removing this H^+ causes further dissociation of the acetic acid into H^+ and acetate ions according to Equation 2-4; the OH^- of the titrant consumes the H^+ as they dissociate; the change in pH resulting from this interaction follows the curve shown in Figure 2-1. Obviously, as titrant is added, the amount of acetate ions in the system increases; indeed the amount of acetate formed is almost the same as the amount of OH⁻ added. The addition of 10 ml of the titrant yields a 0.1 M sodium acetate solution with the pH expected of a solution of this salt (i.e., pH $\cong$ 8.4).

The shape of the acetic acid titration curve is complex; it should be noted that the steepest portion of the curve occurs at the point where approximately 5 ml of titrant has been added and that, for a pH unit on either side of the steepest point on the curve, it is almost a linear curve. This fact and other aspects of the shape of this curve will be discussed in further detail when buffers are considered.

The titration of boric acid yields a titration curve with a shape similar to that of acetic acid, in terms of the steep linear portion of the curve. That the boric acid is graphed to the right of the acetic acid relates solely to the fact that boric acid is a much weaker acid than acetic acid. Moreover any number of weak acids could have been chosen, and the shape of the titration curves of identical concentration solutions would be similar to that of the acetic acid in the same sense that the boric acid curve is, but it would lie either to the right or the left of the acetic acid curve depending on the strength of the weak acid. The fact that the slope of the linear region of these titration curves is similar for each weak acid relates to the fact that each has the capacity to donate the same amount of protons as NaOH is added; of course the X axis is a log function which means that the amount of free $[H^+]$ change differs for each weak acid, but the amount of free H^+ is negligible compared to the total amount of H^+ donated by the HA during the titration. The similar slopes of the linear regions of these curves make it possible to derive equations for buffers of general applicability, i.e., not restricted to one specific buffer system.

It should be mentioned that this similarity in titration curves of weak acids is obvious only if the H^+ concentration is expressed in logarithmic form, such as pH. A plot of $[H^+]$ on the X axis in place of pH or other log expression of H^+ concentration yields curves that would not show similarities between acetic and boric acids. This is an example of the importance of being able to think about H^+ not only in terms of concentration but also in terms of pH.

BUFFERS

Aqueous solutions in which the addition of acid or alkali causes far less of a change in pH than is the case with pure water are called buffer solutions. The solutes responsible for the resistance to pH change are referred to as buffers or buffer systems. The simplest kind of buffer system is a combination of a weak acid and its conjugate base (i.e., a solution containing HA and A^-). Of course buffer systems can also be quite complex with the involvement of many different weak acids and their conjugate bases (blood is an example of a complex buffer system).

The simple case, a weak acid and its conjugate base (i.e., a single buffer pair), is worth considering in detail because it permits ready understanding of the principles involved. Such a system resists pH change because when H^+ ions are added most react with the conjugate base A^- to form HA with only a few remaining in solution as H^+; thus little change in pH occurs. Similarily when OH^- ions are added, the H^+ reacting with OH^- to form H_2O is rapidly generated by the dissociation of HA to H^+ and A^-, thus preventing any marked fall in $[H^+]$ or rise in pH as a result of the OH^- addition. It should be emphasized, however, that although buffer pairs minimize pH change resulting from addition of H^+ or OH^- to aqueous solutions, some change of pH does occur. The amount of pH change that will occur can be calculated and the following section lays the groundwork for such quantitative assessments.

Henderson-Hasselbalch Equation

A most useful tool for such quantitative considerations is the Henderson-Hasselbalch equation which can be derived from Equation 2-13. The first step in the derivation is the rearrangement of Equation 2-13 in such a way that $[H^+]$ becomes the sole term on the left hand side of the equation as in Equation 2-19:

$$[H^+] = K' \frac{[HA]}{[A^-]} \qquad (2\text{-}19)$$

The next step is to transform Equation 2-19 into a log form as in Equation 2-20:

$$\log [H^+] = \log K' + \log \frac{[HA]}{[A^-]} \qquad (2\text{-}20)$$

When each term of Equation 2-20 is multiplied by -1, Equation 2-21 is obtained.

$$-\log [H^+] = -\log K' + \log \frac{[A^-]}{[HA]} \qquad (2\text{-}21)$$

The last term on the right hand side of Equation 2-21 is not preceded by a negative sign because $-\log \frac{[HA]}{[A^-]} = +\log \frac{[A^-]}{[HA]}$ i.e., $-\log \frac{[HA]}{[A^-]}$, obtained in multiplying Equation 2-20 by -1, is reexpressed as $+\log \frac{[A^-]}{[HA]}$. As defined in Chapter 1, pH $\simeq -\log [H^+]$, and in an analogous way a term called pK' is defined as follows: $pK' = -\log K'$. By replacing $-\log [H^+]$ with pH, and $-\log K'$ with pK', the Henderson-Hasselbalch equation (Equation 2-22) is obtained:

$$pH = pK' + \log \frac{[A^-]}{[HA]} \qquad (2\text{-}22)$$

The application of the Henderson-Hasselbalch equation to buffers can probably best be approached initially in relation to the titration curve for acetic acid in Figure 2-1. When 5 ml of the NaOH titrant was added to the acetic acid solution a pH of approximately 4.7 was obtained; at this point about one-half of the acetic acid had been converted to acetate ion and one-half remained as undissociated acetic acid. Therefore in this particular case there is an equimolar mixture of acetic acid and its conjugate base, acetate ion. On the basis of the definition of what constitutes a buffer system, clearly at this point in the titration, the system is an acetic acid-acetate buffer system, one which will resist pH change caused by the addition of either H^+ or OH^-.

Moreover the effect of adding H^+ or OH^- are graphically shown by the titration curve in Figure 2-1. Much of this curve can be worked out theoretically, i.e., by pencil-and-paper chemistry, by means of the Henderson-Hasselbalch equation. For instance, the pH of 4.7, when 5 ml NaOH are added, can be calculated by Equation 2-22 on the basis of the following data: the pK' of this acetic acid solution is 4.7; one-half the acetic acid has been converted to acetate ion which means that acetic acid is present at a $\sim$0.05 M concentration and acetate ion is present at a $\sim$0.05 M concentration. Clearly the pH is the only unknown in Equation 2-22 and its value can be readily calculated. There is an assumption made in this calculation, however, namely, that the amount of acetate ion in the solution is equal to the amount of OH^- added. This is not strictly true because there is a very small amount of acetate ion formed from the disssociation of acetic acid in addition to the acetate ion generated by titration with OH^-. The error made is so small that it is sufficiently accurate for most purposes to disregard it and to make the calculation by Equation 2-22 on the assumption that the amount of acetate ion is equal to the amount of added OH^-.

The Henderson-Hasselbalch equation can be used just as effectively for calculating the pH of a mixture of acetic acid and sodium acetate formed by adding the acid and salt as such instead of generating the acetate by titrating with OH^-. To generalize further, the Henderson-Hasselbalch equation is useful in calculating the pH of a buffer system whenever significant quantities of both the undissociated weak acid and its conjugate base are present; therefore, most of the titration curve for weak acids, such as acetic acid and boric acid, can be predicted by pencil-and-paper calculation.

However, at the extremes of the titration curve, where the HA term is either very large or very small relative to A^-, the Henderson-Hasselbalch equation does not provide a good approximation of the pH. If only HA has been added, obviously all A^- is derived from dissociation of the weak acid, but none of the dissociation is caused by titration with OH^- or by the direct addition of A^- as a salt. Since use of Equation 2-22 is based on the assumption that most of the A^- was caused by the addition of either OH^- or the A^- salt, the Henderson-Hasselbalch equation is not of use in that region of the titration curve (the extreme left in Fig. 2-1) where this assumption does not hold. At the other extreme (the extreme right in Fig. 2-1) the value of HA approaches zero; mathematically, it is not meaningful to divide by zero, and, practically, when the value of the HA term is very small, the error in using the Henderson-Hasselbalch equation is too great even for obtaining an approximate value of pH.

Along a different track, it should be noted that, when equimolar concentrations of HA and A⁻ are present, the pH value is numerically equal to the pK′ (i.e., the log of 1 is zero, so that the second term of the right hand side of the Henderson-Hasselbalch equation becomes zero when the concentration of the weak acid and its conjugate base are the same). Because of this, the Henderson-Hasselbalch equation also serves a useful role in determining the pK′ of a buffer pair.

A QUANTITATIVE VIEW OF BUFFERS

The groundwork has been laid for the consideration of buffers quantitatively. For this, the titration curve for acetic acid in Figure 2-1 is a good starting point. When the NaOH titrant is added to the aqueous acetic acid solution, there is at the start a marked increase in pH, but when a pH of ∼3.7 is reached the changes in pH occur more slowly as more titrant NaOH is added. This resistance to pH change increases until a pH ∼ 4.7 (the pK′ of acetic acid is ∼4.7) is reached at which point there is maximum resistance to pH change when more titrant is added. Further addition of titrant continues to involve only small changes in pH, but the resistance to pH change becomes less as the pH continues to rise. Above a pH ∼ 5.7 there is little resistance to pH change when more titrant is added.

If instead of 1 liter of a 0.1 M acetic acid being titrated with NaOH, a 0.1 M sodium acetate solution (pH ≅ 8.4) were titrated with 10 M HCl as the titrant, the same curve that occurs with NaOH titration of acetic acid will be obtained, but it will be traced in the reverse direction. That is, as the HCl titrant is added a rapid change in pH will occur until a pH ≅ 5.7 is reached; further addition of HCl titrant causes the pH to fall slowly with maximum resistance to pH change occurring at pH ∼ 4.7. Further addition of HCl titrant meets considerable but lessening resistance to pH change until a pH of ≅ 3.7 is reached. Further addition of HCl titrant causes the pH to fall markedly.

The experiment just presented on the titration of acetic acid or sodium acetate describes the buffering action of the acetic acid-acetate buffer pair in quantitative fashion. First, it establishes that a buffer pair is able to resist pH change most when the pH is of a value equal to the pK′ of the weak acid. Secondly, it establishes that there is a good buffer capacity, although not a maximal one, for one pH unit on either side of the pK′ value. In the particular case cited, acetic acid-acetate functions as a buffer in the pH range of 3.7 to 5.7 with a maximum efficiency at a pH ≅ 4.7. To generalize, in the selection of a buffer system for a particular job, a buffer pair should be chosen with a pK' of its weak acid within a pH unit of the pH desired for the system.

Of course, buffering power also relates to the concentration of the buffer components. Obviously a buffer pair present at a 1 mM concentration has far less buffer power than the same system at a 10 mM concentration— one tenth as much if both are tested at the same pH.

In considering buffers quantitatively, the concept of *buffer value* is of use. The *buffer value* of a solution is defined in terms of the slope of the titration curve of that buffer system at a given pH. Obviously the slope of the curve of any buffer system at a given concentration is greatest at the pK'. If the tangent of the curve at the pK' is drawn, then the concept of *maximum buffer value* of a buffer pair is expressed in terms of the amount of acid or alkali required to change the pH by one unit if buffering power were expressed throughout the pH change by the slope of this tangent. Obviously the slope does change during a pH change of one unit. Nevertheless, the *maximum buffer value* is a useful one for the theoretical consideration of buffers quantitatively. Using this concept and also taking into consideration the importance of concentration in determining buffer power, we can define the term *maximum molar buffer value* as the buffer value of one liter of a 1 M solution in terms of the number of moles of acid that would be required to change the pH by one unit on the basis of the slope of the tangent to the pK' point of the titration curve. This value is 0.575 moles of acid per liter of 1 M buffer for any ionizing group undergoing titration. For example, phosphoric acid (H_3PO_4) has three groups that can release or accept H^+, each with a different pK', and for each of these groups a *maximum molar buffer value* of 0.575 moles is obtained.

Obviously if the pK' of a buffer system is known, it is possible to say at what pH region this buffer will be effective at a glance. However, such knowledge is not obtained at a glance, if [H^+] is used instead of pH. This is another example of the pH concept being more useful than [H^+]. This is not to imply, however, that pH is always more useful than [H^+] because, as stated earlier, there are may circumstances in which it is far more convenient for the physician and physiologist to think in terms of [H^+] rather than pH. Obviously both pH and the [H^+] are important concepts and each is to be used in the circumstances where it best meets the needs.

PROBLEMS

1. To make a liter of a 0.1 M acetate buffer solution of pH 4.5 at 25°C from glacial acetic acid and sodium acetate, how many millimoles of each of this buffer pair is required? The pK' of acetic acid under these conditions is 4.7.

2. How many ml of 1 N NaOH would be needed to change the pH of the buffer solution prepared in Problem 1 from a pH of 4.5 to a pH of 5.0?

3. If the pH of an aqueous solution is 5.0, what is the approximate OH⁻ concentration of this solution?

REFERENCES

For Modern Concepts of Acids and Bases:

Pearson, R. G.: Hard and soft acids and bases, HSAB. Part I Fundamental principles. J. Chem. Ed. *45*: 581–587 (1968).
Pearson, R. G.: Hard and soft acids and bases, HSAB. Part II Underlying theories. J. Chem. Ed. *45*: 643–648 (1968).

For Titration and Theory of Buffers:

Bull, H. B.: An Introduction to Physical Biochemistry, Chapter 5. F. A. Davis Co., Philadelphia, 1964.

For Use of Arrhenius Concept:

MacConnachie, N. F.: An old-fashioned approach to acid-base balance. Amer. J. Medicine. *49*: 504–518 (1970).

ACIDS, BASES, AND BUFFERS IN MAMMALS

PHYSIOLOGIC APPLICATION OF BRØNSTED-LOWRY SYSTEM

Compounds such as glucose, ethyl alcohol, and triglyceride are Brønsted-Lowry acids because under certain circumstances these compounds can function as proton donors. However, they do not donate protons to the water or buffer systems of mammals, and therefore from a physiologic point of view it is not useful to consider them as acids—indeed it is confusing to do so. In order to exclude such compounds from the classification of acids, however, it is necessary to restrict the domain of the Brønsted-Lowry system of acids for biomedical purposes to include only those compounds that can serve as proton donors in the living organism. Such a definition excludes glucose, ethyl alcohol, triglyceride, and many other compounds but includes acetic acid, hydrochloric acid, lactic acid, and similar compounds that can donate protons to body water and buffer systems. We shall call the acids in this restricted domain physiological Brønsted-Lowry acids, and when henceforth the term acid is used in this book it will refer exclusively to such compounds.

A similar restriction must be placed on the domain of Brønsted-Lowry bases for biomedical purposes. The chemist properly considers chloride, sulfate, and similar ions to be Brønsted-Lowry bases since there are circumstances where these compounds serve as proton acceptors. However, under the conditions found in the mammalian organism, such compounds do not function as proton acceptors and therefore they should be excluded from any biomedical consideration of bases. The domain of the Brønsted-Lowry system for bases is restricted for biomedical purposes to the physiological Brønsted-Lowry bases defined as compounds that can function as proton acceptors in the body. This definition excludes chloride and sulfate as physiological Brønsted-Lowry bases but includes HCO_3^-, $HPO_4^=$, and many other compounds to be discussed later.

A discussion of the phosphoric acid-phosphate system illustrates the importance of restricting the domain of the Brønsted-Lowry system for physiological purposes. From a chemist's viewpoint $H_2PO_4^-$ can function either as an acid or a base. For it to function significantly as a base, however, the pH of the system must range between 1 and 4. Since the pH of body fluids is around 7, clearly $H_2PO_4^-$ cannot significantly function physiologically as a base.* It does, however, function as an acid in the body, since in the pH range of 7 it can donate protons to the body water and other proton acceptors. From a chemist's point of view, $HPO_4^=$ can also function as both an acid and a base. For this compound to function as an acid, the pH of the system must be above 8; clearly therefore $HPO_4^=$ under physiological circumstances does not function significantly as an acid. It does, however, serve as a base because at the pH of the body fluids it is a most effective acceptor of protons. The difficulties that can be encountered in dealing with biomedical problems when the domain of the Brønsted-Lowry system is not restricted is made evident in the recent paper of Camien and co-workers cited at the end of this chapter.

Two general classes of physiological Brønsted-Lowry acids can be defined on the basis of certain chemical and physical properties and on the differences in the way in which the acid is processed by the mammal. The first class is composed of a single member, carbonic acid (H_2CO_3), which can dissociate into H^+ and HCO_3^- and is also in equilibrium with CO_2 and H_2O; this system is expressed symbolically by Equation 3-1:

$$CO_2 + H_2O \rightleftharpoons H_2CO_3 \rightleftharpoons H^+ + HCO_3^- \qquad (3-1)$$

Carbonic acid is uniquely different from other physiological Brønsted-Lowry acids for the following reasons: (1) CO_2 readily permeates cell membranes and thus rapidly influences intracellular $[H^+]$; (2) large amounts of CO_2 are continuously being produced by metabolic processes; and (3) CO_2 is continuously being eliminated from the body by the lungs.

The second class of physiological Brønsted-Lowry acids is the noncarbonic acids composed of all acids in the body other than carbonic acid. Typical examples of this class are H_2SO_4, lactic acid, and acetoacetic acid; these acids are neither volatile nor in equilibrium with a volatile component such as CO_2 and therefore are often called fixed acids. The noncarbonic acids (or fixed acids) are processed by the living organism in a fashion distinctly different from that of carbonic acid. Moreover since most are almost fully dissociated at body pH and since H^+ and most anions permeate

---

* An exception to this statement is the fluid in the gastric lumen where the pH is normally <4.

most cell membranes only with difficulty, the noncarbonic acids cannot rapidly influence intracellular [H$^+$] unless generated within that cell.

BUFFERS OF PHYSIOLOGICAL IMPORTANCE

From a physiological point of view the buffer systems of the body can also be divided into two classes. One class, composed of a single member called the bicarbonate buffer system, involves H_2CO_3 and HCO_3^- as its buffer pair; the other class comprises all other buffers which will collectively be called nonbicarbonate buffer systems. The bicarbonate buffer system is a most important one for the mammal. However, because it functions somewhat differently from the buffers discussed so far, a detailed description of the properties of the bicarbonate buffer system will be delayed until Chapter 4. In this chapter brief mention will often be made of the bicarbonate buffer system, but no attempt will be made to describe its chemical mechanism.

The body contains many compounds and functional groups in macromolecules which are capable of functioning as buffers. However, most of these compounds and groups do not function as significant buffer systems in the living organism because the pK' of the weak acid member of the buffer pair is too far removed from the value of 7 to enable the buffer system to readily donate or accept protons in the physiological pH range of 7, i.e., such compounds or groups under physiological circumstances are either almost completely protonated or completely dissociated. There are two general kinds of compounds other than the bicarbonate system that function as buffers in the physiological pH range: the proteins and phosphate compounds. An example of a phosphate buffer is the inorganic orthophosphate system comprised of the dihydrogenphosphate ($H_2PO_4^-$) monovalent anion and monohydrogenphosphate ($HPO_4^=$) divalent anion which function as shown in Equation 3-2:

$$H_2PO_4^- \rightleftharpoons HPO_4^= + H^+ \tag{3-2}$$

The pK' of the acid form, i.e., the $H_2PO_4^-$, is 6.8; clearly with a pK' of this value this buffer can readily function in the maintenance of the pH of the body fluids. In the pH range of approximately 7, the acid member $H_2PO_4^-$ of this buffer system readily donates protons, and the conjugate-base form $HPO_4^=$ readily accepts them. Many of the organic phosphate compounds found in the body have pK' values within one-half of a pH unit of 7.0; these compounds like the inorganic orthophosphate buffer system function well as buffers under physiological conditions as shown in Equation 3-3:

$$
\begin{array}{ccc}
\overset{\displaystyle O}{\underset{\displaystyle OH}{\overset{\|}{A-O-P-O^-}}} & \rightleftharpoons & \overset{\displaystyle O}{\underset{\displaystyle O_-}{\overset{\|}{A-O-P-O^-}}} + H^+
\end{array}
\tag{3-3}
$$

In Equation 3-3, the symbol A refers to an organic structure to which phosphate is monoesterified; the acid form of the organic phosphate buffer pair is on the left and the conjugate base on the right. Obviously with the pK′ values quoted above, these compounds can readily serve as proton donors or proton acceptors in the physiological pH range. Specific examples of such organic phosphates are glucose-1-phosphate, the nucleoside monophosphates such as adenosine monophosphate, the nucleoside diphosphates such as adenosine diphosphate, and the nucleoside triphosphates such as adenosine triphosphate.

In the case of the proteins, the most important group in these macromolecules that can function as a buffer system in the physiological pH range is the imidazolium group of the histidine residues. Equation 3-4 depicts the way in which the imidazolium group functions as a buffer:

$$\cdots CO—NH—CH—CO—NH\cdots \quad \cdots CO—NH—CH—CO—NH\cdots$$

$$
\begin{array}{ccc}
 & \underset{\displaystyle \underset{+HN \quad\quad NH}{\overset{\displaystyle HC == C}{|}}}{\underset{CH}{}} & \rightleftharpoons \quad \underset{\displaystyle \underset{N \quad\quad NH}{\overset{\displaystyle HC == C}{|}}}{\underset{CH}{}} + H^+
\end{array}
$$

$$(3\text{-}4)$$

The dotted lines refer to the peptide chain of the protein. The pK′ of the acid form on the left hand side of the equation is usually in the neighborhood of 7, making it an excellent buffer system in the physiological pH range. However, it must be recognized that the pK′ of any group of an amino acid residue of a protein is affected by the structure of the protein and by the neighboring amino acid residues. In the case of the imidazolium group of histidine residues the pK′ of different imidazolium groups have been found to vary from 5.3 to 8.3, depending on the protein structure in the immediate environment.

Most other potential buffering groups in the proteins do not function at the physiological pH range. For example, the α carboxyl of the C-terminal amino acid residue, the β carboxyl of aspartate residues, and the γ carboxyl of glutamate residues have pK′ values less than 5. The phenolic group of the tyrosine residues, the sulfhydryl of the cysteine residues, the ε amino of lysine residues, and the guanidinium of arginine residues have pK′ values above 8. Indeed the only other group capable of buffering at the pH of body fluid is the α amino group of N-terminal amino acid residues, a group that has a pK′ range from 7.8 to 10.6. Since there is usually only one N-terminal amino acid residue per protein molecule and the pK′ of the group is at best somewhat high to be optimal for buffering under physiological circumstances, it adds little to what histidine residues provide in regard to the physiological buffering power of proteins in body fluids.

THE ISOHYDRIC PRINCIPLE

The body fluids of mammals do not contain just one buffer pair or system but several different buffer pairs. If a fluid compartment is homogeneous in regard to the distribution of the several buffer pairs (as with different buffer pairs in water solution contained in a beaker) then the detailed knowledge of a single buffer pair (i.e., its pK' and the concentration of the weak acid and conjugate base) permits ready assessment of both the [H+] of the homogeneous solution and the molar ratio of weak acid to conjugate base of all other buffer pairs in the system provided the pK' of each of the buffer pairs is known. The isohydric principle is the formal name used to designate the fact that all buffer pairs in homogeneous solution are in equilibrium with the same [H+]. For example, if there are five different buffer pairs dissolved in a well-mixed solution, the entire system can be quantitatively assessed by the following equation which is an algebraic statement of the isohydric principle:

$$[H^+] = K_1' \frac{[HA_1]}{[A_1^-]} = K_2' \frac{[HA_2]}{[A_2^-]} = K_3' \frac{[HA_3]}{[A_3^-]} = K_4' \frac{[HA_4]}{[A_4^-]} = K_5' \frac{[HA_5]}{[A_5^-]}$$

$$(3\text{-}5)$$

The numerical subscripts in Equation 3-5 refer to a specific buffer pair. The isohydric principle can be expressed in terms of pH by Equation 3-6:

$$pH = pK_1' + \log \frac{[A_1^-]}{[HA_1]} = pK_2' + \log \frac{[A_2^-]}{[HA_2]} = pK_3' + \log \frac{[A_3^-]}{[HA_3]}$$

$$= pK_4' + \log \frac{[A_4^-]}{[HA_4]} = pK_5' + \log \frac{[A_5^-]}{[HA_5]} \qquad (3\text{-}6)$$

For biomedical purposes the isohydric principle makes it possible to learn much about the acid-base chemistry of plasma by the analysis of one of the buffer pairs, the single buffer pair usually measured being the bicarbonate buffer system. Of course of all body fluids, the plasma is particularly well suited for the use of the isohydric principle since it is a homogeneous solution in which the various buffer pairs are uniformly distributed.

BLOOD BUFFERS

Blood is the tissue which the physician samples for the analysis of the acid-base state of his patient. Therefore, the buffer system of blood should be emphasized for study both because of the physiological importance of blood buffers and because of their relationship to clinical chemistry.

From the point of view of buffering, blood is a two compartment system, composed of plasma (making up 55 per cent of its volume) and cells (comprising the other 45 per cent). Although the blood contains cells other than erythrocytes (or red blood cells), only the erythrocytes, because of their great abundance relative to any other cell type, need be considered in regard to the cellular compartment of the blood buffer system.

Plasma is a salt solution (ionic strength of about 0.15) which contains about 7 per cent protein. The plasma contains three kinds of buffers: the bicarbonate buffer system, the $H_2PO_4^-/HPO_4^=$ buffer system, and the buffering of certain groups in the plasma proteins (albumin and the various classes of globulins). It should be recognized that these proteins contain a series of different buffer systems since the imidazole group of each histidine residue has a different pK', the magnitude of which depends on the influence of other amino-acid residues or other molecular entities in the neighborhood of that histidine residue. Therefore, the plasma proteins provide a broad spectrum of histidine residue buffer pairs with a spread of pK' values ranging from about 5.5 to 8.5. For this reason, plasma proteins provide a continuous series of buffer pairs throughout the pH region encountered in plasma in health and disease.

This very complicated plasma protein buffer system is often written as a simple pair, with the acid form expressed as HPr and the conjugate-base form as Pr⁻. This formulation, although a useful shorthand, is misleading if it is not realized that at the pH of plasma the proteins are polyanions and their buffering primarily involves either the addition of a proton to the imidazole group when an acid is being buffered or the removal of a proton from the imidazole group when an alkali is being buffered. Obviously when a proton is added to the imidazole groups the plasma protein becomes less polyanionic (thus the rationale for the shorthand HPr), and when alkali is added the plasma proteins become more polyanionic (and thus the symbol Pr⁻).

When noncarbonic (or fixed) acids are added to normal plasma more than 75 per cent of the buffering capacity of plasma resides in the bicarbonate buffer system which will be described in detail in Chapter 4. Most of the rest of the buffering of noncarbonic acid involves the plasma protein buffers with a small contribution from the $H_2PO_4^-/HPO_4^=$ system. The reason the phosphate is not important quantitatively as a plasma buffer relates to its low concentration in plasma; certainly with a pK' of 6.8, the $H_2PO_4^-/HPO_4^=$ system would be an important one quantitatively if present in an appreciable concentration. A quantitative description of buffering by plasma must be delayed until Chapter 4.

Although plasma is not an important site of buffering of carbonic acid, the plasma proteins and, to a lesser extent, the inorganic phosphate buffer systems have some capacity in this regard. The bicarbonate buffer system plays no role in the buffering of carbonic acid; the reason for this will be presented in Chapter 4. It should be noted that while plasma has a considerable buffer capacity for the buffering of noncarbonic acids (fixed acids), it has much less capacity to buffer carbonic acid.

The isohydric principle is most useful when considering plasma since all the buffer pairs are in equilibrium with each other and with the $[H^+]$ of the plasma (see Equation 3-5). It should be noted that plasma is an homogeneous buffer compartment.

Although the bicarbonate buffer is also quantitatively important in the red blood cell compartment, hemoglobin is quantitatively even more important. Hemoglobin is present in these cells at a very high concentration, and of the 540 amino acid residues in its molecule, 36 are histidine residues. Moreover the N-terminal valine residue amino groups have a pK' of about 7.8 and thus can also serve in a buffer capacity at the pH range found physiologically in red blood cells. Indeed this combination of histidine residues and N-terminal valine amino residues cause the hemoglobin titration curve to be almost totally linear in the pH range of 7 to 8. Besides the bicarbonate and hemoglobin buffer systems, the erythrocytes also contain a significant amount of organic phosphate which functions as a buffer in this pH range. The buffering capacity of the organic phosphates, however, is quantitatively small compared to the bicarbonate and hemoglobin buffer systems in the red blood cell.

The hemoglobin buffer system is often written in the following symbolism: HHb for the weak-acid form of nonoxygenated hemoglobin; Hb^- for the conjugate-base form of the nonoxygenated hemoglobin; $HHbO_2$ for the weak-acid form of the oxygenated hemoglobin; and HbO_2^- for the conjugate-base form of the oxygenated hemoglobin. Clearly as in the case of plasma proteins, this symbolism is merely a shorthand for the fact that the protonated imidazole groups of histidine residues and amino groups of N-terminal valine residues lead to a lowering of the net negative charge on the hemoglobin polyanion molecule, while dissociation of protons from these groups leads to an increased negative charge. The oxygenated hemoglobin is treated as a separate buffer system from the nonoxygenated hemoglobin because there is a change in the pK' value for some of the histidine residues when hemoglobin changes from a nonoxygenated to an oxygenated molecule. This phenomenon is of great physiological importance and will be discussed at length in Chapter 5.

When noncarbonic acids (fixed acids) are being buffered, more than 60 per cent of the buffer capacity of the red cells resides in the hemoglobin buffer systems, and more than 30 per cent in the bicarbonate buffer system. The remaining buffer capacity for noncarbonic acid in the red cells resides with the organic phosphate buffers. The red blood cell bicarbonate buffer, as with the plasma bicarbonate buffer, plays no role in the buffering of carbonic acid. However, the red blood cells have a great capacity to buffer carbonic acid because of the high concentration of hemoglobin buffer systems. Indeed more than 90 per cent of the blood's capacity to buffer carbonic acid resides in the hemoglobin buffer system.

Physiologically the blood, in regard to its buffering activity, is usually considered as a whole rather than in terms of each separate compartment.

However, the red blood cell buffer system and the plasma can not be linked together by use of the isohydric principle because each is a separate compartment, so that the blood cannot be considered as a single homogeneous system. Indeed proof that the isohydric principle is not valid is provided by the data on the pH value of each compartment; the plasma normally has a pH of about 7.4, while the pH of the red blood cells is approximately 7.2.

Although the plasma and the red blood cell compartments cannot be simply linked by use of the isohydric principle, they are not independent of each other. Certain components of their complex buffer systems, especially the bicarbonate buffer system, can readily traverse the membrane of the erythrocyte. The difference in pH between red blood cells and the plasma relates to the following two factors: (1) the erythrocyte is continuously producing lactic acid by anaerobic glycolysis, and (2) many ions such as the plasma protein polyanions, the hemoglobin polyanions, K^+ and Na^+ either do not cross the plasma membranes or are impeded in doing so, causing a Donnan-membrane system to play an important role in the distribution of the diffusible ions including H^+. Of these two factors, the latter is the predominant reason for the pH being lower in red blood cells than in plasma.

All this notwithstanding, the blood is usually considered by physiologists and physicians as approximating a single buffering compartment. Moreover, the red blood cell compartment, although clearly intracellular in anatomical terms, is thought of in the physiological context of body buffers as a part of the extracellular buffering system. This is a valid approximation because of its very rapid interaction with plasma. Indeed henceforth in this book, blood as a whole will be considered to be part of the extracellular buffering system of the body; the reader must adapt to this useful but flagrant misuse of anatomical fact.

BUFFERS OF THE INTERSTITIAL FLUID

The bicarbonate buffer system is the major one in the interstitial fluid (including the lymph). Indeed, on the basis of the Donnan-membrane equilibrium effect being the dominant factor in the interaction between the interstitial fluid and plasma at the capillary endothelium, the concentration of HCO_3^- in the interstitial fluids would be expected to be about 5 per cent higher in interstitial fluid than in plasma. Interstitial fluid also contains the $H_2PO_4^-/HPO_4^=$ buffer pair, but as in plasma its concentration is too low for it to be a quantitatively important buffering system. The concentration of the plasma proteins in interstitial fluid varies from one region of the body to another, but by and large interstitial fluid has a very low concentration of these and other proteins. Therefore, the histidine residues of proteins do not serve a quantitatively important role as buffer components of the interstitial fluid.

The interstitial fluid per unit of volume has almost as great a capacity to buffer noncarbonic acids as does the plasma. In both plasma and interstitial fluid the bicarbonate buffer system is the most important one for the buffering of noncarbonic acid, and the concentration of HCO_3^- is at least as great in the interstitial fluid as in plasma. Moreover since the total volume of interstitial fluid is $3\frac{1}{2}$ to 4 times that of the plasma, it is obvious that the capacity of the total interstitial fluid to buffer noncarbonic acid is great. Indeed the total capacity of man's interstitial fluid compartment to buffer noncarbonic acid is considerably more than the capacity of his total blood volume to buffer these acids.

On the other hand, the interstitial fluid has little capacity to buffer carbonic acid. The small amounts of phosphate buffer and protein in interstitial fluid are the only buffers in this compartment capable of buffering carbonic acid. As stated before, the bicarbonate buffer system cannot function as a buffer for carbonic acid.

If the Donnan-membrane equilibrium effect at the capillary endothelial membrane which separates the interstitial fluid from the plasma is considered to be the dominant factor influencing the $[H^+]$ of the interstitial fluid, then the interstitial fluid should have a H^+ concentration about 5 per cent lower than the plasma, e.g., if plasma has a concentration of 40 nmoles per liter, the interstitial fluid would be expected to have one of about 38 nmoles per liter. However, the interstitial fluid is also in contact with the multitude of metabolizing cells of the body which are producing both carbonic and noncarbonic acids. These acids diffuse through the interstitial fluid on their way to the plasma and thus influence the interstitial fluid so as to increase its H^+ concentration beyond that predicted when the Donnan-membrane equilibrium between plasma and interstitial fluid was considered to be the sole factor involved. The H^+ concentration in the interstitial fluid undoubtedly varies relative to the region of the body and the physiologic state of the organism, but it probably has a value similar to that of the plasma in any steady-state or near steady-state condition.

DYNAMICS OF THE INTERACTION BETWEEN PLASMA AND INTERSTITIAL FLUID

In the near steady-state circumstance of a normal individual or a patient suffering from chronic disease, such as a chronic respiratory disease causing a steady-state disturbance in acid-base balance, it is to be anticipated that the $[H^+]$ of the interstitial fluid will be closely related to that of the plasma, and the difference between them in H^+ and HCO_3^- concentrations can be approximated by assuming the Donnan-membrane equilibrium to be the dominant factor. However, when a mammal is responding to acute deviations from its normal state—deviation, either induced in the laboratory or occurring naturally—this simple relationship between plasma and interstitial fluid no longer holds.

For instance, if a large amount of noncarbonic (fixed) acid is infused intravenously, it is very quickly buffered by the blood buffer system. Transiently the [H^+] of plasma will be considerably higher than that of the interstitial fluid because of the time that it takes to mix the infused acid with the large interstitial fluid volume which contains much of the bicarbonate buffer used to buffer noncarbonic acid. For instance, following the intravenous administration of a large load of HCl it takes about one-half hour for the added H^+ to be buffered throughout the extracellular fluid space, and it is not until this time has elapsed that the H^+ concentration difference between plasma and interstitial fluid again relates primarily to the Donnan-membrane effect. Obviously during the mixing period, the H^+ added by infusion of the HCl is buffered by an ever larger buffer pool, i.e., more and more of the buffering system of the interstitial fluid is brought into play. Thus the plasma [H^+] will decrease with a time course related to the gradual mixing of the infused acid throughout the buffer space of the extracellular fluid. Acute administration of NaOH will lead to similar transient responses, which of course differ in the direction of the [H^+] change in both the interstitial fluid and plasma compartments from that seen following the administration of HCl.

An infusion of carbonic acid into the vascular compartment can also be readily accomplished by increasing the CO_2 content of the inspired air (the chemical mechanism will be discussed in Chapter 4), and the time course can be followed by maintaining the inspired air at this elevated CO_2 content. Even though CO_2 has very great capacity to diffuse into all areas of the body, it takes more than one-half hour before there is mixing of both CO_2 and H^+ to the point that the differences in [H^+] between interstitial fluid and plasma again relate primarily to the Donnan-membrane equilibrium. In the first two minutes or so the [H^+] of the plasma has not risen very markedly because the blood (particularly the red blood cells) has great capacity to buffer carbonic acid. Moreover, in this short period of time there has been little intermixing of the large interstitial compartment, which has little or no buffering capacity for carbonic acid, with the blood compartment. However, by 30 minutes to an hour the effects of this change in CO_2 tension of inspired air will be found to have caused a much greater reduction in the pH of plasma than noted at 2 minutes. The primary reason for this is that during the 30 minutes both H^+ and buffer pairs that can cross the capillarly endothelium readily have mixed to the point where the buffering capacity for carbonic acid of the entire extracellular fluid space is the prime factor setting the [H^+] of the plasma. Since the entire extracellular fluid space has a much lower buffer capacity per unit volume for carbonic acid than does blood, the [H^+] will increase in the plasma as equilibration between the weakly buffered interstitial fluid and the strongly buffered plasma becomes complete. Certainly by one hour after such an increase in carbonic acid, the intermixing between these compartments should be totally complete.

From the foregoing discussion, it is obvious that sampling of blood for analysis of an acid base disturbance can be misleading if the sampling occurs in a transient period following the administration of acid or alkali intravenously to a man. The physician should be fully cognizant of this when dealing with patients suffering from acute deviations in the input of either noncarbonic or carbonic acids or alkali, to the blood. Moreover, when the therapeutic intravenous use of solutions for acid-base abnormalities are underway, the physician should be fully aware of these transients as he monitors the patient.

INTRACELLULAR pH AND BUFFER SYSTEMS

Knowledge of acid-base regulation in the intracellular fluid compartment has not advanced rapidly primarily because of the conceptual and technical difficulties involved in the consideration of the intracellular pH. Although various methods have been used to measure intracellular pH, none is entirely satisfactory. Besides the technical difficulties in making pH measurements within the cells, the fact that a cell is a highly compartmentalized structure with a large series of subcompartments separated from each other by internal membrane systems and having potentially different pH values has suggested to some that the concept of an intracellular pH may neither be valid nor useful. Others, including the authors, feel it is useful to calculate pH values within cells, regarding them in rough terms as "the average pH." However, the reader should realize that although this concept of an "average intracellular pH" may have some use, it is not rigorously defined, nor is it based on a totally sound theoretical foundation.

A pH meter fitted with microelectrodes which can be inserted into cells has been a popular way of measuring the pH of muscle cells. At first glance it might seem that such direct measurement of pH is the method of choice; however, there are several technical drawbacks to such measurements. These difficulties involve the uncertainties regarding the liquid junction potentials of electrodes in a complex suspension like cytoplasm and the uncertainties inherent in placing an electrode into a nonhomogeneous, markedly compartmentalized system. Nevertheless, many investigators using this method have obtained a pH value in skeletal muscle cells of approximately 7.0. However, there also have been reports of pH measurements within the muscle cells indicating a pH of 6.0.

Another popular method for estimating intracellular pH is based on the distribution of a weak acid between the intracellular fluid and the extracellular fluid. The weak acid that recently has been most used is 5,5-dimethyl-2, 4-oxazolidinedione (DMO). The pK' of this weak acid at the ionic strength and temperature of the body fluids is 6.13; the equilibrium between the weak-acid and conjugate-base forms is shown in Equation 3-7:

$$
\begin{array}{ccc}
\underset{\displaystyle |}{CH_3} & & \underset{\displaystyle |}{CH_3} \\
CH_3-\underset{\displaystyle \underset{O}{\overset{|}{\;}}}{C}\underset{\underset{HDMO}{\text{(weak acid)}}}{\text{———}}\underset{\displaystyle \underset{NH}{\overset{|}{\;}}}{C}=O & \rightleftharpoons & CH_3-\overset{|}{C}\text{———}\overset{\|}{C}-O^- \qquad + H^+
\end{array}
$$

HDMO
(weak acid)   DMO⁻ (conjugate base)

$$(3\text{-}7)$$

This method is based upon the reasonable assumption that the un-dissociated weak-acid form readily penetrates the cell membranes and enters the cell where it dissociates according to the intracellular pH. It is also assumed that the conjugate-base form of this weak acid system cannot permeate the cell membrane in either direction. By using DMO labeled with radio-isotopes the amount of material used can be so low that the H^+ produced by its dissociation is negligible. When equilibrium has occurred it is assumed that the concentration of the undissociated acid form is the same in the intracellular fluid as in the extracellular fluid. By measuring the concentration of the total DMO—both the undissociated weak acid (HDMO) and its conjugate-base form (DMO⁻)—within the intracellular fluid, the intracellular pH can be calculated by Equation 3-8:

$$
pH = 6.13 + \log \frac{\underset{\text{(within cell)}}{[\text{total DMO}]} - \underset{\text{(within cell)}}{[\text{HDMO}]}}{\underset{\text{(within cell)}}{[\text{HDMO}]}} \tag{3-8}
$$

Since [total DMO] − [HDMO] is equal to DMO⁻, the numerator of Equation 3-8 is the conjugate-base concentration just as in all other uses of the Henderson-Hasselbalch equation. The [HDMO] within the cell is assumed to be the same as the [HDMO] in the extracellular fluid, a value which can be readily assessed.

Recently the DMO method has been used to measure the pH of several different rat tissues. Rat skeletal muscle was found to have a pH of 6.93, heart muscle of 7.04, and liver a pH of 7.23. If the DMO method provides an "average intracellular pH," this pH is somewhat different in different kinds of mammalian cells. Moreover, as will be discussed later, changing physiological conditions will cause the intracellular pH of tissues to change.

Clearly from most of the work on the "average intracellular pH," it can be estimated that the intracellular H^+ concentration is about 100 nmoles/L. This seems particularly well established for skeletal muscle, a tissue comprising a large fraction of the cytoplasmic mass of a man. If H^+ were distributed passively between the extracellular fluid and the cytoplasm of the muscle cell, then the Nernst equation, Equation 3-9, could be used to calculate the intracellular H^+ concentration.

$$E_m = 61 \log \frac{[H^+]_0}{[H^+]_i} \qquad (3\text{-}9)$$

E_m denotes membrane potential in mV, $[H^+]_0$ the concentration of H^+ in extracellular fluid, and $[H^+]_i$ the concentration of H^+ in intracellular fluid. On the basis of the membrane potential of skeletal muscle cells being 90 mV negative on the intracellular surface, as compared to the extracellular surface, and the interstitial fluid H^+ concentration being approximately 40 nmoles/L, if, on this basis, H^+ were distributed in equilibrium in accord with this electrochemical gradient, the H^+ concentration within the muscle cell should be about 1000 nmoles/L. If most of the experimental data are correct, the H^+ concentration in the muscle cells is about 1/10th this concentration. It must therefore be postulated that the skeletal muscle has some kind of active transport system which sets the intracellular $[H^+]$ at a lower concentration than would be established by physico-chemical forces alone. More work is needed to say which of the other mammalian tissues must also have similar active transport systems.

The intracellular fluid does not have a high concentration of the bicarbonate buffer system but does contain significant amounts of non-bicarbonate buffer systems. The large amount of intracellular protein, with its content of histidine residues, and the large amount of organic phosphate compounds within cells are the quantitatively significant non-bicarbonate buffer systems. Thus the intracellular fluid is able to buffer effectively both noncarbonic and carbonic acids as well as alkali.

DYNAMICS OF INTERACTION BETWEEN INTRACELLULAR FLUID AND EXTRACELLULAR FLUID

Normally the intracellular pH is held at a steady-state concentration by the action of buffers and active transport processes yet to be clearly defined. In chronic deviations from normal, such as certain chronic respiratory diseases, the intracellular pH may or may not differ from normal and can be held at a new steady-state concentration by these means. The response of the intracellular pH to acute changes in extracellular pH, although not fully worked out, is a time-dependent event and the usual time course of this response should be generally understood by physicians and experimental physiologists.

If a noncarbonic acid, such as HCl, is administered intravenously for an hour or so, much of this infused H^+ will eventually leave the extracellular space, enter the cells and be buffered there, but the time course of entry of H^+ into the cell is slow. For some time most of the administered H^+ remains in the extracellular fluid where it is buffered by the extracellular fluid buffers (red blood cell buffers included of course). Then gradually over a period of several hours much of the H^+ from this intravenous load of noncarbonic acid enters the intracellular compartment. In this way, the buffering capacity of the intracellular compartment is slowly utilized to buffer the infused HCl.

Intravenously administered alkali is handled in a similar way, i.e., fluxes between the cells and the extracellular fluid slowly make the intracellular buffers available for buffering the alkali. Also certain cells, most probably skeletal muscle cells, generate lactic acid and release it to the extracellular fluid, thus serving to minimize the rise in extracellular fluid pH caused by the administration of an alkali.

Increasing the pCO_2 of the alveolar air results in the addition of carbonic acid to the blood (see Chapter 4 for details). The administration of carbonic acid differs from that of noncarbonic acid in several ways. Carbon dioxide readily permeates all regions of the body. However, a mixing time of one-half to 1 hour or so is needed before the CO_2 (and thus H_2CO_3) is distributed by cardiovascular system to all tissues at an equal concentration. At this point, CO_2 has formed H_2CO_3 in the intracellular as well as the extracellular compartment, the intracellular nonbicarbonate buffer systems being used to buffer much of the H^+ generated intracellularly from H_2CO_3. There then occurs a slow interplay between the extracellular fluid and the intracellular fluid which takes several hours to come to completion but which leads to the buffering by intracellular buffers of some of the H^+ which accumulated in the extracellular fluid.

In the case of a rapidly occurring deficit in carbonic acid, such as occurs during voluntary hyperventilation, there is again a half-hour period or so before the elimination of CO_2 from all compartments has led to a nearly equal concentration of carbonic acid throughout the body. A rise in pH occurs both intracellularly and extracellularly. Then a slow interaction between the extracellular fluid and the intracellular fluid compartments occurs which modifies the pH change in the extracellular fluid by slowly utilizing the intracellular fluid buffers to readjust extracellular fluid pH. Moreover the skeletal muscle responds to the change in pH caused by the hyperventilation by producing rather large quantities of lactic acid which of course help mitigate the rise in pH in all compartments of the body.

PROBLEMS

1. The pK' of the $H_2PO_4^-$ in plasma at $98°F$ is 6.8. At the normal pH of 7.40 of plasma, what is the molar ratio of $HPO_4^=/H_2PO_4^-$?

2. How many milliliters of 0.1 N HCl solution will be required to titrate 100 ml of 0.02 M $HPO_4^=/H_2PO_4^-$ buffer solution from a pH of 7.4 to a pH of 6.8? In this problem assume a pK' for $H_2PO_4^-$ of 6.8 and the temperature is $98°F$.

3. If when the plasma is at pH 7.4 and 98°F the molar ratio of a particular histidine residue of plasma albumin is 1.2/1 in favor of the nonprotonated imidazole ring, what is the pK' of the imidazole group of that particular histidine residue?

REFERENCES

For Physiologic Application of Brønsted-Lowry System:

Camien, M. N., Simmons, O. H., and Gonick, H. C.: A critical reappraisal of "acid-base" balance. Amer. J. clin. Nutr. *22*:786–793 (1969).

For Buffers of Physiological Importance:

Williams, V. R., and Williams, H. B.: Basic Physical Chemistry for the Life Sciences, Chapter 4. W. H. Freeman and Co., San Francisco, 1967.

For the Isohydric Principle:

Winters, R. W., Engel, K., and Dell, R. B.: Acid-Base Physiology in Medicine. The London Co., Westlake, Ohio, 1967.

For Blood Buffers:

Calvey, T. N.: Intracellular pH in mature and immature rabbit red cells. Quant. J. Exp. Physiol. *55*, 238–252 (1970).
Winters, R. W., and Dell, R. B.: Regulation of acid-base equilibrium. *In* Yamamoto, W., and Brobeck, J. (eds.): Physiological Controls and Regulations. W. B. Saunders Co., Philadelphia, 1965.

For Interstitial Fluid Buffers and the Dynamics of Interaction Between Plasma and Interstitial Fluid:

Woodbury, W. H.: Regulation of pH. *In* Ruch, T. C., and Patton, H. D. (eds.): Physiology and Biophysics. W. B. Saunders Co., Philadelphia, 1965.

For Intracellular pH and Buffers:

Butler, T. C., Waddell, W. J., and Poole, D. T.: The pH of intracellular water. Ann. N. Y. Acad. Sci. *133*:73–77 (1966).
Carter, N. W., Rector, F. C. Jr., Canpion, D. S., and Seldin, D. W.: Measurement of intracellular pH of skeletal muscle with pH – sensitive glass microelectrodes. J. Clin. Invest., *46*:920–933 (1967).
González, N. C., and Genge, O. A.: Effect of changes in CO_2 tension upon hydrogen ion activity of the skeletal muscle of the dog. Life Sciences *9*: part II, 961–966 (1970).
Sanslone, W. R., and Muntwyler, E.: Intracellular pH changes associated with protein and potassium deficiency. Metabolism *17*:1084–1093 (1968).
Siesjö, B. K., and Pontén, U.: Intracellular pH—true parameter or misnomer? Ann. N. Y. Acad. Sci. *133*:78–86 (1966).
Tushan, F. S., Bromberg, P. A., Shively, A. G., and Robin, E. D.: Intracellular pH and electrolyte metabolism in chronic stable hypercapnia. Arch. intern. Med. *125*:967–974 (1970).
Waddell, W. J., and Bates, R. G.: Intracellular pH. Physiol. Rev. *49*:285–329 (1969).
Walker, W. D., Goodwin, F. J., and Cohen, R. D.: Mean intracellular hydrogen ion activity in the whole body, liver, heart and skeletal muscle of the rat. Clin. Sci. *36*:409–417 (1969).

4

CARBONIC ACID, THE BICARBONATE BUFFER SYSTEM, AND A QUANTITATIVE VIEW OF BUFFERING IN BODY FLUIDS

PROPERTIES OF CARBONIC ACID

Carbonic acid (H_2CO_3) has certain chemical properties which differ from the noncarbonic acids. Moreover the physiological processing of H_2CO_3 differs from that of the noncarbonic acids. The role of these unique chemical and physiological properties of H_2CO_3 in acid-base physiology will be explored in this chapter.

The dissociation of H_2CO_3 is summarized in Equation 4-1:

$$H_2CO_3 \rightleftharpoons H^+ + HCO_3^- \rightleftharpoons H^+ + CO_3^= \qquad (4\text{-}1)$$

The value of the pK' in aqueous solution at 38°C for the dissociation of

H_2CO_3 to HCO_3^- (bicarbonate) and H^+ is approximately 3.8 and that for the dissociation of HCO_3^- to H^+ and $CO_3^=$ (carbonate) is about 9.8. Our interest will be focused on the dissociation of H_2CO_3 to H^+ and HCO_3^-; little mention will be made concerning the dissociation of HCO_3^- to H^+ and $CO_3^=$ since at the pH of the body fluids, little $CO_3^=$ is present.

On the basis of Equation 4-1 it would appear that H_2CO_3, at least from a chemical point of view, is a typical weak acid. However, Equation 4-1 does not describe the only reaction that H_2CO_3 undergoes; H_2CO_3 is also in equilibrium with the CO_2 dissolved in aqueous solutions as shown in Equation 4-2:

$$CO_2 + H_2O \rightleftharpoons H_2CO_3 \qquad (4\text{-}2)$$

At the temperature and ionic strength of the body fluids, the equilibrium between H_2CO_3 and CO_2, as shown by Equation 4-2, lies far to the left; specifically there are about 500 CO_2 molecules in solution for every H_2CO_3 molecule. That H_2CO_3 is in equilibrium with CO_2 has important consequences for the mammalian organism. For instance, CO_2 readily diffuses across cell membranes. As a result, any changes in H_2CO_3 concentration and, therefore, in CO_2 concentration in a fluid compartment of the body will soon influence every other fluid compartment of the body.

Furthermore the CO_2 dissolved in an aqueous solution is in equilibrium with the CO_2 in any gas phase in contact with that solution. In the body, therefore, it is the gas phase in the alveoli of the lungs that is in equilibrium with the CO_2 dissolved in the body fluids. This relationship between dissolved CO_2 and gaseous CO_2 is symbolized by Equation 4-3:

$$\underset{\text{in gas phase}}{CO_2} \rightleftharpoons \underset{\text{in aqueous solution}}{CO_2} \qquad (4\text{-}3)$$

The concentration of a particular gas in a mixture of gases is usually expressed in terms of its partial pressure in the gas mixture. The partial pressure of CO_2, designated pCO_2, is usually expressed in mm Hg. The concentration of a gas dissolved in an aqueous solution is proportional to the partial pressure of that gas in the gas phase; this relationship is called Henry's law which quite accurately defines the situation for pressures up to 760 mm Hg. The actual amount of a gas dissolved per unit volume of aqueous solution is defined by the following parameters: the chemical nature of the gas, the temperature of the solution, the solutes dissolved in the solution, and the partial pressure of the gas. For blood plasma at 38°C the amount of CO_2 dissolved is expressed by Equation 4-4:

$$[CO_2] = 0.03\ pCO_2 \qquad (4\text{-}4)$$

In this equation, $[CO_2]$ relates to millimoles dissolved CO_2 per liter of plasma rather than the more usual meaning of the square bracket of moles per liter; pCO_2 is in mm Hg and 0.03 is the proportionality constant for plasma at 38°C. If the pCO_2 of the alveolar air is 40 mm Hg (the usually stated normal value) then the concentration of CO_2 in the plasma is 1.2 mmoles/L. The concentration of a dissolved gas in the aqueous solution is sometimes expressed in terms of mm Hg, the so-called tension of the dissolved gas. By this system, the plasma in equilibrium with the gaseous pCO_2 of 40 mm Hg would have a CO_2 tension of 40 mm Hg. Although this shorthand is quite useful in respiratory physiology it does not provide the insight for acid-base physiology that the conversion of pCO_2 into mmoles/L does.

What has been said about H_2CO_3 up to this point can be summarized by Equation 4-5:

$$CO_2 \rightleftharpoons CO_2 \quad + \quad H_2O \rightleftharpoons H_2CO_3 \rightleftharpoons H^+ + HCO_3^- \qquad (4\text{-}5)$$
$$\text{in gas phase} \quad \text{in aqueous solution}$$

It should be noted that the almost trace level of H_2CO_3 in the body fluids is in equilibrium with a much larger pool of the dissolved CO_2. Moreover since dissolved CO_2 and H_2CO_3 are in equilibrium, the total concentration of the trace H_2CO_3 and the large CO_2 reservoir pool can be combined in a so-called total "carbonic acid pool" (i.e., CO_2 plus H_2CO_3), and the equilibrium constant for this combined "carbonic acid pool" and its dissociation products can be calculated by Equation 4-6:

$$K = \frac{[H^+][HCO_3^-]}{[CO_2] + [H_2CO_3]} \qquad (4\text{-}6)$$
$$\text{dissolved}$$

The re-expression of Equation 4-6 in terms of pH and pK yields Equation 4-7:

$$pK = pH + \log \frac{([CO_2] + [H_2CO_3])}{[HCO_3^-]}^{\text{dissolved}} \qquad (4\text{-}7)$$

The pK′ of the system expressed in Equation 4-7 for plasma at physiological pH and 38°C is 6.1. As will be discussed later it is this pK′ value which is of significance when considering the bicarbonate buffer system because the entire pool of potential H_2CO_3 (both H_2CO_3 and CO_2) is available to the organism for buffering purposes.

THE BICARBONATE BUFFER SYSTEM

Since the pK′ of the total "carbonic acid pool" is 6.1, it is to be expected that at a pH range around 6.0 this bicarbonate-carbonic acid system will function effectively as a buffer. It is also evident that, in the analysis of this buffer system, the total reservoir of carbonic acid, i.e., both dissolved CO_2 and H_2CO_3, termed in this text the total "carbonic acid pool," should be considered to be the acid component of the bicarbonate buffer system, and HCO_3^- the conjugate base. The Henderson-Hasselbalch equation modified as shown in Equation 4-8 can be used for quantitative analysis of the bicarbonate buffer system:

$$pH = pK' + \log \frac{[HCO_3^-]}{[CO_2 + H_2CO_3]} \qquad (4\text{-}8)$$

Since the pCO_2 is a physiologic measurement readily obtained experimentally, the denominator of the log term in Equation 4-8 can be replaced by the pCO_2 expressed in mm Hg multiplied by the factor 0.03, provided the temperature is 38°C and the ionic strength is 0.15. Since this calculation expresses the total "carbonic acid pool" in terms of millimoles per liter, the concentration of HCO_3^- in the numerator must also be expressed in millimolarity. At this ionic strength and temperature, the pK' of the total "carbonic acid pool" is 6.1. Combining all of these data, the following convenient working equation for physiologic studies can be set forth:

$$pH = 6.1 + \log \frac{[HCO_3^-]}{0.03 \ pCO_2} \qquad (4\text{-}9)$$

It should be again emphasized that HCO_3^- in the numerator is expressed in millimolarity rather than molarity as usually designated by [ ].

Normally the plasma of arterial blood has a pH of about 7.40 and a pCO_2 of about 40 mm Hg. Therefore, by Equation 4-9, it can be calculated that the bicarbonate concentration in such plasma is about 24 mmoles of plasma which is indeed the case. From what was said about buffers in Chapter 2, the fact that the pK' of the bicarbonate buffer system is 6.1 indicates that it should not be an important buffer for the maintenance of the normal plasma pH since its pK' is more than a unit less than the normal pH of plasma. However, because of certain physiological characteristics of the mammal and the chemical nature of CO_2, the bicarbonate buffer is a very effective one for the maintenance of the plasma pH. The reason for this effectiveness can be made evident by a very simple *in vitro* system which defines the way the bicarbonate buffer functions in closed and open systems.

If a beaker contains 1 L of 0.15 M NaCl solution, 24 mmoles of bicarbonate and 1.2 mmoles total "carbonic acid pool" (i.e., primarily dissolved CO_2) the concentration of the two components of the bicarbonate buffer system is similar to that of plasma under *in vivo* circumstances. If the beaker is sealed in some hypothetical manner so that this solution is not in contact with a gaseous phase, an ideal closed system has been set up as depicted in Figure 4-1A; the pH of the solution is 7.40 $\left(\text{i.e., pH} = 6.1 + \log \frac{24}{1.2} = 7.4\right)$. If, while the closed system is maintained, 12 mmoles of HCl (Fig. 4-1B) are added, the following reaction will occur:

$$Na^+ + HCO_3^- + H^+ + Cl^- \rightarrow CO_2 + H_2O + Na^+ + Cl^-$$

As a result of this reaction, approximately 12 mmoles of HCO_3^- will have been converted to CO_2 and the pH of the solution will have shifted to 6.06 $\left(\text{i.e., pH} = 6.1 + \log \frac{12}{13.2} = 6.06\right)$. Clearly as predicted in Chapter 2, the bicarbonate buffer system is a poor one for maintaining the pH in the range of 7.4.

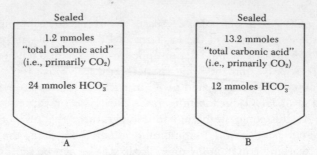

Figure 4-1. Closed system. Beaker containing 1 L of a 0.15 M NaCl solution (38°C) in which are dissolved 24 mmoles NaHCO₃ and 1.2 mmoles "total carbonic acid." *A* refers to conditions before HCl is added, and *B* refers to conditions after 12 mmoles of HCl were added.

An open system can be set up in the beaker containing 1 L of the same solution as the closed system by bubbling through this solution a gas containing CO_2 at 40 mm Hg partial pressure. If the pCO_2 of this gas is maintained at 40 mm Hg, the amount of CO_2 dissolved in solution can be maintained at 1.2 mmoles per liter come what may. This system is depicted in Figure 4-2A, and the pH of the solution is also 7.40. When 12 mmoles of HCl (Fig. 4-2B) are added to this open system, 12 mmoles of HCO_3^- are again converted to CO_2, but, unlike the closed system, this CO_2 does not remain in solution but is carried away by the gas stream which continues then to maintain a pCO_2 of 40 mm Hg so that the solution remains at 1.2 mmoles of dissolved CO_2 per liter. For this reason the addition of the HCl will cause much less of a change in pH than in the closed system, as shown

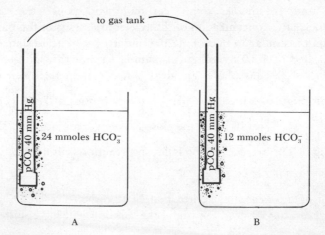

Figure 4-2. Open system. Beaker contains 1 L of 0.15 M NaCl solution (38°C) in which is dissolved 24 mmoles of NaHCO₃. The solution is being continuously gassed with a gas mixture containing 40 mm Hg pCO_2. *A* refers to conditions before HCl is added, and *B* refers to conditions after 12 mmoles HCl were added.

by the calculation of the pH of the solution following the addition of the HCl:

$$pH = 6.1 + \log \frac{12}{1.2} = 7.1$$

This calculation makes clear that in the open system the bicarbonate buffer system is a fairly effective buffer of noncarbonic acid in the pH range of 7.4. Moreover, this open system not only functions by eliminating CO_2 when noncarbonic acid is added but also serves as a source of CO_2 for buffering added alkali. For instance 1.2 mmoles of dissolved CO_2 in a closed system of Figure 4-1A would not be able to effectively buffer 12 mmoles of NaOH which would consume all carbonic acid available by the time 10 per cent of the OH^- had been added; the pH upon addition of the alkali would therefore increase to 12 or so. In an open system, however, where the gas source serving as provider of CO_2 maintains CO_2 at 1.2 mmoles/L (Fig. 4-2), 12 mmoles of NaOH is effectively buffered as follows:

$$CO_2 + H_2O \rightarrow H_2CO_3$$
$$H_2CO_3 + Na^+ + OH^- \rightarrow Na^+ + HCO_3^- + H_2O$$

Summary: $\quad CO_2 + Na^+ + OH^- \rightarrow Na^+ + HCO_3^-$

The pH would increase only to 7.58 $\left(pH = 6.1 + \log \dfrac{36}{1.2} = 7.58\right)$.

The body is an open system much as the one depicted in Figure 4-2. The metabolic activities provide a continuous source of CO_2, i.e., they generate CO_2. The ventilatory system continuously eliminates CO_2 as regulated by neural mechanisms. The body can therefore regulate the CO_2 concentration in the fluids to meet ever changing needs by controlling the rate at which the lung eliminates CO_2.

Many years ago Van Slyke developed the so-called pH-HCO_3^- diagram, and Davenport (see references) subsequently utilized it most effectively for teaching purposes. Unfortunately the pH-HCO_3^- diagram is not greatly used in clinical work. Nevertheless because of the insight it provides for the beginner, its concepts will be developed in this chapter. The *in vitro* open system just described is such a simple one that it permits an easy introduction to the pH-HCO_3^- diagram. In the pH-HCO_3^- diagram, pH is plotted along the X axis and the HCO_3^- concentration in mmoles/L along the Y axis as in Figure 4-3. The solution in the beaker in Figure 4-2A has 24 mEq HCO_3^-/L, a pH of 7.40, and a pCO_2 of 40 mm Hg (38°C); this set of data is designated by point A in the pH-HCO_3^- diagram. Since this system contains only bicarbonate buffer, it can be used to study the buffering of noncarbonic acids by the bicarbonate buffer alone.

As HCl, a typical noncarbonic acid, is added to this open system, HCO_3^- accepts the proton and is converted to CO_2 which is eliminated by the gas stream, i.e., the pCO_2 is maintained at 40 mm Hg by the gas stream. Since the amount of HCO_3^- consumed is approximately equal to the

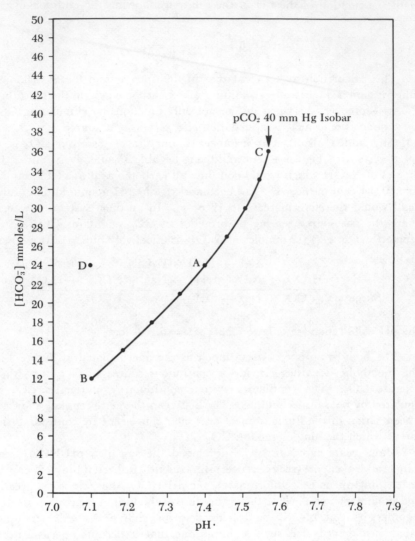

Figure 4-3. pH-HCO_3^- diagram for open system of Figure 4-2 gassed with pCO_2 40 mm Hg with exception of point D where pCO_2 is 80 mm Hg.

millimoles of HCl added, the change in HCO_3^- concentration is readily calculated, i.e., addition of 1 mmole HCl decreases HCO_3^- to 23 mmoles/L, 2 mmoles HCl decreases HCO_3^- to 22 mmoles/L, and so forth. Since the pCO_2 is held constant, the change in pH with these additions can be readily calculated by Equation 4-9 and the relationship between pH and [HCO_3^-] at pCO_2 40 mm Hg can be plotted on the pH-HCO_3^- diagram.

    The curve A to B is a plot of the relationship of HCO_3^- concentrations and pH at pCO_2 40 mm Hg as HCl is added to the beaker. Point B is reached when 12 mmoles of HCl is added causing the loss of about 12 mmoles of HCO_3^-. This curve is called the pCO_2 40 mm Hg isobar because the pCO_2 is maintained by the open system at 40 mm Hg. This graph can be extended from point A to point C by adding NaOH stepwise to the open system ($CO_2 + Na^+ + OH^- \rightarrow NaHCO_3$). For every millimole of NaOH added, approximately 1 mmole of $NaHCO_3$ will be formed by reacting with CO_2, but of course the pCO_2 remains at 40 mm Hg because the gas stream continues to supply it. At point C, 12 mmoles of NaOH have been added.

That the bicarbonate buffer system can not effectively buffer carbonic acid can also be demonstrated in this open system containing only bicarbonate buffer. If the pCO_2 in the gas stream in Figure 4-2A is increased from 40 mm Hg to 80 mm Hg the total "carbonic acid pool" will be increased from 1.2 to 2.4 mmoles/L. This change in pCO_2 causes a fall in pH and almost no change in HCO_3^- concentration. The extent of these changes can be readily quantitated by use of Equation 4-9. The pH will fall to 7.1 $\left(pH = 6.1 + \log \dfrac{24}{2.4} = 7.1 \right)$. Of course since an increase in the pCO_2 from 40 to 80 mm Hg causes the H_2CO_3 concentration to double, there will in fact be some increase in $[HCO_3^-]$ because of the dissociation of H_2CO_3 ($H_2CO_3 \rightarrow H^+ + HCO_3^-$), the increase in $[HCO_3^-]$ being equal to the increase in $[H^+]$. The increase in H^+ concentration (estimated from the fall in pH from 7.4 to 7.1) is 40 nmoles/L and therefore the increase in HCO_3^- will be 40 nmoles or 40×10^{-6} mEq/L. This would mean that the HCO_3^- concentration in the beaker will have increased from 24.00000 to 24.00004 mmoles/L, a change that cannot be measured chemically, nor accurately plotted on the graph in Figure 4-3, nor is it physiologically important and thus can be neglected.

The set of data pCO_2 80 mm Hg, pH 7.1, $[H^+]$ 80 nM, HCO_3^- 24 mEq/L are represented by point D on the graph. Therefore an increase in the "carbonic acid pool" of 1.2 mmoles/L caused the generation of only 40 nmoles of H^+, none of which is buffered; this should be compared to the fact that for the noncarbonic acid HCl to cause the same fall in pH in this system, 12,000,000 nmoles of H^+ need to be added, of which 11,999,960 are buffered by the bicarbonate buffer with 40 nmoles remaining free in solution. Clearly the bicarbonate buffer system is an effective buffer in an open system for noncarbonic acids but does not function as a buffer for carbonic acid.

In contrast to the bicarbonate buffer system which can only buffer noncarbonic acids, the nonbicarbonate buffer systems can buffer both noncarbonic and carbonic acids. The nonbicarbonate buffer systems can also be studied in the open system shown in Figure 4-2A. To consider nonbicarbonate buffers, an amount X of a nonbicarbonate buffer (designated HBuf for the acid component and Buf^- for the conjugate-base component) will be added to the solution in Figure 4-2A which already contains the bicarbonate buffer. Let's consider this HBuf-Buf^- to be a protein containing a series of histidine residues with pK' values ranging from 7.0 to 7.8. Since the solution in Figure 4-2 now contains both bicarbonate and nonbicarbonate buffer systems, to study only the nonbicarbonate system, it is necessary to titrate with carbonic acid (it must be recalled that carbonic acid is not buffered by bicarbonate buffer). The pH-HCO_3^- diagram in Figure 4-4 will be used to analyze the buffering of H_2CO_3 by the HBuf:Buf$^-$ system.

Point A again represents the initial condition of pH 7.4, pCO_2 40 mm Hg, HCO_3^- concentration of 24 mmoles/L (and in this case in addition the

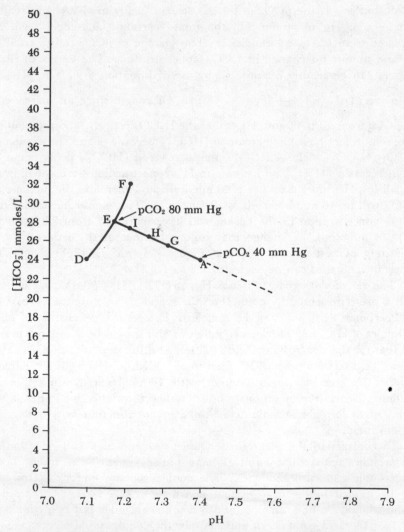

Figure 4-4. pH-HCO_3^- diagram for open system of Figure 4-2 containing nonbicarbonate buffer (HBuf:Buf$^-$). The pCO_2 in open system is varied between 40 mm Hg and 80 mm Hg. Hatched line refers to pH and HCO_3^- changes that might be expected if pCO_2 is reduced to levels less than 40 mm Hg.

HBuf:Buf⁻ buffer system). If in this case the pCO_2 is increased to 80 mm Hg (i.e., an increase in total "carbonic acid pool" of 1.2 mmoles/L), there will be a considerable increase in [HCO_3^-] since Buf⁻ will buffer much of the H⁺ formed from H_2CO_3 as follows: $H_2CO_3 + Buf^- \rightarrow HCO_3^- + HBuf$. In the presence of the X amount of this HBuf:Buf⁻ buffer system let us assume that the [HCO_3^-] increases to 28 mEq/L and the pH falls to 7.17 when the pCO_2 is increased from 40 mm Hg to 80 mm Hg. This set of data pH 7.17, pCO_2 80 mm Hg, HCO_3^- concentration of 28 mEq/L are plotted as point E in Figure 4-4. Since for every HCO_3^- formed from H_2CO_3 an H⁺ must also have formed, the Buf⁻ must have accepted about 4 mmoles of H⁺ (specifically 3,999,972 nmoles of H⁺ were accepted by the nonbicarbonate buffers and 28 nmoles of H⁺ remained in free solution). Note that since the increase in amount of H⁺ in free solution is only 28 nmoles, it can be neglected when calculating the amount of H⁺ accepted by Buf⁻ (i.e., since 4×10^6 nmoles of H⁺ were generated because of the increase in carbonic acid concentration, it is clear that just about all of the H⁺ so generated was accepted by the nonbicarbonate buffer system).

Obviously HBuf:Buf⁻ is an effective buffer for H_2CO_3 as can be seen at a glance by comparing point E in the figure with point D obtained when the pCO_2 is increased to 80 mm Hg in the absence of a nonbicarbonate buffer system. If HBuf:Buf⁻ buffer is increased from amount X to an amount $2X$ in the solution in Figure 4-2A, then approximately twice as much H⁺ from H_2CO_3 will be accepted by Buf⁻ when the pCO_2 is lifted to 80 mm Hg. Point F represents this set of data, i.e., pCO_2 80 mm Hg, HCO_3^- 32 mmoles/L, and pH of 7.22.

By comparing points D, E, and F it is readily apparent that the more HBuf:Buf⁻ present, the more the [HCO_3^-] increases and the less the pH falls with an increase in pCO_2 to 80 mm Hg. It also happens that the curve connecting points D, E, and F is a pCO_2 80 mm Hg isobar, i.e., the pCO_2 has a value of 80 mm Hg at any point on the curve. Obviously such isobars can be constructed for any desired pCO_2, and henceforth the term pCO_2 isobar will be used without further explanation.

Returning to the solution in Figure 4-2 where only X amount of the HBuf:Buf⁻ was added as graphically described in Figure 4-4, the effect of increasing the pCO_2 stepwise from 40 to 80 mm Hg will now be considered; point G, H, and I represent a pCO_2 of 53, 60, and 70 mm Hg respectively. There is almost a straight line relationship between the increase in [HCO_3^-] and fall in pH as the HBuf:Buf⁻ system is titrated with increasing amounts of H_2CO_3 (this is the case because the titration is taking place within a series of similar but not identical pK′ values of the various histidine residues of the protein comprising the HBuf:Buf⁻ system). As previously stated the amount of H⁺ accepted by Buf⁻ is almost exactly equal to the amount of HCO_3^- generated from the H_2CO_3; therefore the slope of the line A–E represents the buffer power of the X amount of HBuf:Buf⁻.

The buffer line A–E can be continued to the right of point A by lowering the pCO_2 to values less than 40 mm Hg; in this case the HBuf:Buf⁻ system will buffer in the following fashion: $HBuf + HCO_3^- \rightarrow H_2CO_3 + Buf^-$. Obviously as the pCO_2 falls the HCO_3^- will decrease and the pH will increase, but the curve will linearly relate $[HCO_3^-]$ to pH change only when the pH is within the range of the pK' values of the histidine residues of the protein comprising this $HBuf{:}Buf^-$ system.

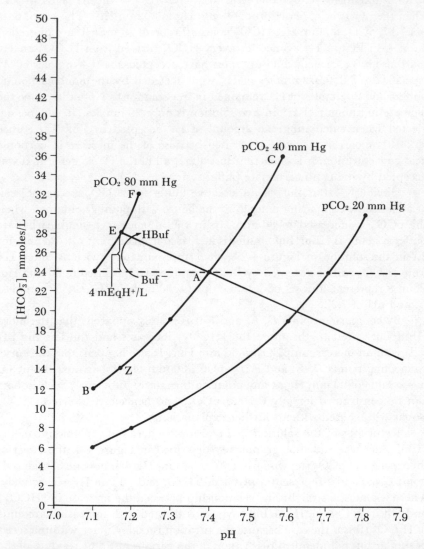

Figure 4-5. pH-HCO_3^- diagram for the analysis of the buffering of HCl by 1 liter of 0.15 M NaCl solution containing 24 mmoles HCO_3^- and X amount of nonbicarbonate buffer ($HBuf{:}Buf^-$). pCO_2 in an open system like that diagrammed in Figure 4-2 is maintained at 40 mm Hg. Line passing through points A and E is the nonbicarbonate buffer curve. Besides the pCO_2 40 mm Hg isobar, the pCO_2 20 mm Hg and 80 mm Hg isobars are also shown, but the discussion of the figure in the text refers to a system maintained solely at pCO_2 40 mm Hg.

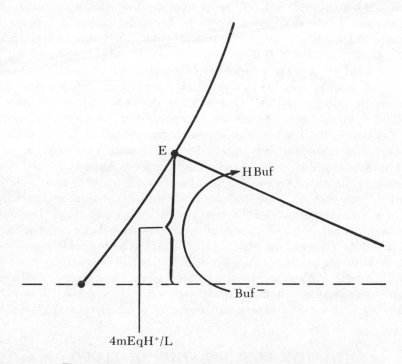

E

HBuf

Buf⁻

$4mEqH^+/L$

Figure 4-5A.   Enlargement of bracket region in Figure 4-5.

Linear relationships between pH and $[HCO_3^-]$, like the line plotted through points E, I, H, G, and A in Figure 4-4, are called nonbicarbonate buffer curves, and the steepness of the line is a quantitative assessment of the amount of nonbicarbonate buffer present in the system. Of course experimentally the slope of the nonbicarbonate buffer curve can only be obtained by titrating with carbonic acid; however, once the slope is known for a system it can be used in the analysis of the buffering of noncarbonic acids by the system. This concept of a *nonbicarbonate buffer curve* and its plot on the pH-HCO_3^- diagram which relate steepness of slope to amount of non-bicarbonate buffer is of great use and should at this time be fixed clearly in mind before reading further.

The effect of adding a noncarbonic acid (e.g., HCl) to the solution in Figure 4-2A containing X amount of HBuf:Buf⁻ can readily be analyzed on the basis of the data and concepts discussed relative to Figures 4-3 and 4-4. This information is combined in Figure 4-5. If starting at point A, under conditions where the pCO_2 is maintained at 40 mm Hg by the open system, enough HCl is added to drop the pH from 7.40 to 7.17, then from the pCO_2 40 mm Hg isobar of Figure 4-5 it is evident that the HCO_3^- concentration will have fallen from 24 to 14 mEq/L, i.e., to point Z. Therefore the bicarbonate buffer system has accepted almost 10 mEq of H^+ from the HCl added to this liter of solution. Moreover from the nonbicarbonate buffer line (i.e., the line connecting points A and E) it can be estimated that the X amount of the nonbicarbonate buffer system (HBuf:Buf$^-$) must have accepted 4 mEq of H^+ from the HCl if the pH has fallen to 7.17. The bracket } is used in Figure 4-5 (see Figure 4-5A for enlarged sketch of bracket region of Figure 4-5) to quantitate the amount of H^+ taken up by Buf$^-$ because with noncarbonic acid the $[HCO_3^-]$ decreases and does not relate to the amount of H^+ taken up by Buf$^-$. Therefore it is inappropriate to relate the buffering action of Buf$^-$ to the Y axis as was done with carbonic acid when the slope of the curve was experimentally established. Therefore 14 mmoles of HCl must have been added to the beaker of which $\frac{2}{7}$ was buffered by the nonbicarbonate buffer system and $\frac{5}{7}$ by the bicarbonate buffer system.

This section should have made clear the role of the bicarbonate buffer system and the importance of the open system in its functioning. It should also have delineated the function of nonbicarbonate buffer systems and established the power of the pH-HCO_3^- diagram as an indicator not only of the total buffering power but also of distribution of buffering activity between the bicarbonate and nonbicarbonate buffer systems. The theory established with the simple *in vitro* buffer systems depicted in Figure 4-2 will now be used to learn about the complex buffer systems of biological materials.

QUANTITATIVE CONSIDERATION OF BLOOD AS A BUFFER SYSTEM

Blood contains both bicarbonate and nonbicarbonate buffer systems. The continuous series of pK' values of the nonbicarbonate buffer systems makes it possible to consider the total set of nonbicarbonate buffer systems as a single theoretical entity in the pH range of 7.0 to 8.0.

An instrument called the tonometer permits the *in vitro* study of blood as a buffer system in what amounts to an open system. The tonometer is a container of large volume into which a small volume of blood is introduced for equilibration with the large gas volume; the tonometer is constructed so that its gas composition can be readily changed to any desired composition. The gas volume of the tonometer is so much larger than the volume of added blood that any gas taken up by or released from the blood to the gas phase will have a negligible effect on the composition of the gas phase. Therefore if the tonometer contains CO_2 at a partial pressure of 40 mm Hg, after equilibration with blood has occurred the pCO_2 of the gas phase will for all practical purposes still be 40 mm Hg.

After equilibration, a sample of blood can be used for determining its "CO_2 content" by the Van Slyke manometric method. Total "CO_2 content" does not refer only to dissolved CO_2 in blood but also to H_2CO_3,

HCO_3^-, and carbamino-CO_2, i.e., to the various chemical forms by which CO_2 is carried in the blood. Indeed HCO_3^- accounts for most of this "CO_2 content" of blood, and therefore the "CO_2 content" relates most closely to the HCO_3^- content of blood. Since the amount of dissolved CO_2 in blood can readily be calculated from the pCO_2 of the gas phase in equilibrium with the blood, the approximate concentration of HCO_3^- in blood can be estimated as follows: The "CO_2 content" — content of dissolved CO_2 = content of HCO_3^-.* These data can then be used in the Henderson-Hasselbalch equation to calculate pH as follows:

$$pH = 6.1 + \log \frac{(CO_2 \text{ content} - \text{dissolved } CO_2)}{(\text{dissolved } CO_2)}$$

The "CO_2 content" and the dissolved CO_2 are usually expressed as mmoles/L of blood. Of course this calculation does not provide information on the pH of either the plasma or the erythrocytes but merely involves a kind of "average pH" for blood based on the $[HCO_3^-]$ and pCO_2; clearly the plasma would have a higher pH and the red blood cells a lower pH than this "average pH" for whole blood.

It is useful to plot the data obtained with blood on a pH-HCO_3^- diagram. For instance point *A* in Figure 4-6 refers to a blood sample containing 24 mmoles of HCO_3^- per liter, a pH of 7.4, and a pCO_2 of 40 mm Hg in the gas phase of the tonometer. The pCO_2 in the gas phase can be raised stepwise above 40 mm Hg or lowered stepwise below 40 mm Hg as desired. The effect of such changes in pCO_2 on pH and $[HCO_3^-]$ are plotted in the diagram by line *B–C*; this is the nonbicarbonate buffer curve for blood, and its slope is a measure of the amount of nonbicarbonate buffer. As with the open systems described earlier, by increasing the pCO_2 the HCO_3^- concentration increases and the pH decreases. The amount of HCO_3^- increase is a measure of the amount of carbonic acid buffered by the nonbicarbonate buffers of blood. As the pCO_2 is decreased below 40 mm Hg the $[HCO_3^-]$ decreases and the pH increases. It should again be emphasized that the slope of line *B–C* in Figure 4-6 is a measure of the nonbicarbonate buffer power of whole blood. The slope of line *B–C* is linear over the pH range of 7.0 to 7.8 because blood contains many different buffer groups with different pK′ values in this buffer region, thus resulting in fairly uniform buffer power in the pH 7.0 to 8.0 range.

The hemoglobin molecule present within the red cell is the major nonbicarbonate buffer involved, but of course the plasma proteins and phosphate compounds are also involved. Most of the buffering of the carbonic acid occurs within the red blood cells for two reasons: (1) the hydration of CO_2 to form H_2CO_3 (which dissociates to H^+ + HCO_3^-) is a slow reaction and the red blood cells contain an enzyme, called carbonic anhydrase, which

---

* The carbamino-CO_2 is ignored in this calculation and for this reason the calculated concentration of HCO_3^- is an approximate one.

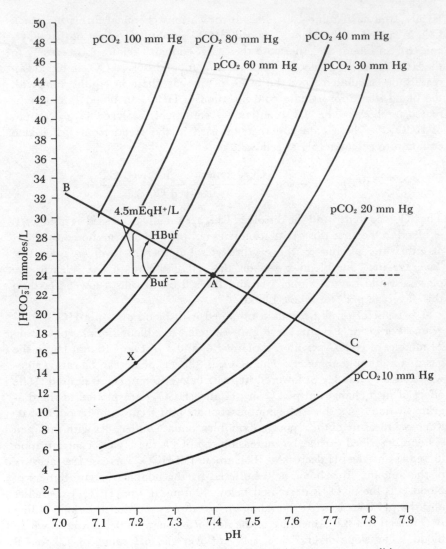

Figure 4-6. pH-HCO$_3^-$ diagram of blood under *in vitro* tonometer conditions.

greatly increases the rate of hydration; (2) The H$^+$ yielded by the dissociation of H$_2$CO$_3$ is primarily buffered by the hemoglobin buffer system which is exclusively in the red blood cells. Although most of the HCO$_3^-$ is generated within the red cell where the H$^+$ from H$_2$CO$_3$ is buffered, most of this HCO$_3^-$ leaves the red blood cells and enters the plasma.

The reason for this migration of HCO$_3^-$ is the Donnan-membrane equilibrium occurring between plasma and red blood cell; since HCO$_3^-$ and Cl$^-$ can readily diffuse across the plasma membrane of the red blood cell, a marked efflux of HCO$_3^-$ to plasma in exchange for Cl$^-$ occurs. Thus not only will the plasma have a higher pH than whole blood but also a higher HCO_3^- concentration. The plasma in equilibrium with red blood cells is called *true* plasma.* The *true* plasma buffer relationships are often presented in a pH-HCO_3^- diagram, but there is little value in doing so now since all it shows is that the true plasma and red blood cells differ quantitatively from the whole blood and that the value for pH and [HCO_3^-] of whole blood lies somewhere between those of the plasma and the red blood cells.

Whole blood is an excellent buffering system for noncarbonic acids not only because of its nonbicarbonate buffers but also because of its bicarbonate buffer system. If HCl is added under conditions where the pCO_2 of the tonometer is maintained at 40 mm Hg, then both the pH and the bicarbonate concentration will fall but the amount of dissolved CO_2 will remain constant. If the HCl is added to blood, starting at point *A* (pH 7.4, HCO_3^- 24 mEq/L) in Figure 4-6 until the pH falls to 7.2, it can be seen from the pCO_2 40 mm Hg isobar that the blood HCO_3^- concentration has fallen to 15 mEq/L, i.e., to point *X*; thus per liter of blood 9 mEq of H^+ were buffered by the bicarbonate buffer system. From the nonbicarbonate buffer curve *B–C* it can be seen that for the pH to fall from 7.4 to 7.2, 4.5 mEq of H^+ from the HCl must have been buffered by the nonbicarbonate buffers per liter of blood (if this point is not obvious, see pages 51 and 52). Therefore in this case, per liter of blood 13.5 mEq of H^+ were buffered (i.e., 13,500,000 nmoles of H^+, a huge amount compared to the increase in free H^+ from 40 nmoles/L to 63 nmoles/L), 4.5/13.5 by nonbicarbonate buffers and 9/13.5 by the bicarbonate buffer.

The same kind of analysis can be made in regard to the addition of NaOH to blood when the pCO_2 is held constant. To quantitatively assess buffering by blood, it is necessary to know whether carbonic or noncarbonic acids or alkali is involved and whether the pCO_2 remains fixed at 40 mm Hg or whether it changes. Because of these complex relationships it was necessary to discuss blood buffering in qualitative terms only in Chapter 3; to discuss it quantitatively the bicarbonate buffer functioning as an open system must be considered. The problem becomes even more complex when the fact that the mammal can adaptively modify the alveolar pCO_2 is also taken into account.

QUANTITATIVE EVALUATION OF BUFFERING UNDER *IN VIVO* CONDITIONS

On the basis of our knowledge of buffering by whole blood as just discussed and the fact that interstitial fluid has almost no nonbicarbonate buffer but about the same concentration of bicarbonate buffer as true

---

* When plasma is separated from red blood cells and studied by itself in a tonometer it is called *separated* plasma. It has little capacity to buffer H_2CO_3 and is of little physiological significance.

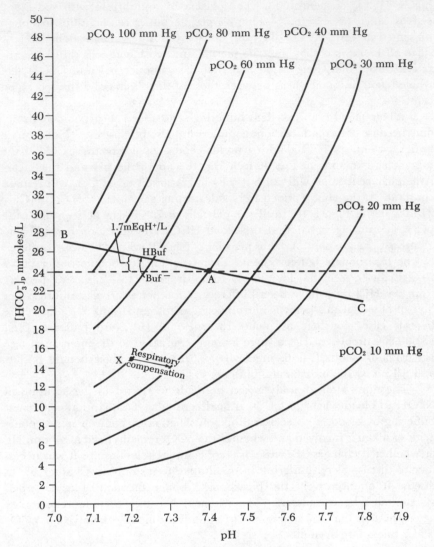

Figure 4-7. pH-HCO$_3^-$ diagram of the "total extracellular fluid." What is meant by "total extracellular fluid" is defined in the text.

plasma, it is possible to develop a model for considering in quantitative terms the buffering activity of the extracellular fluids. This model describes the extracellular fluid as an homogeneous one, i.e., the buffer systems in the erythrocytes, plasma, and interstitial fluid are considered to be homogeneously distributed rather than compartmentalized.

The pH-HCO$_3^-$ diagram shown in Figure 4-7 is used to quantitatively consider this homogeneous model system which shall be designated "total extracellular fluid." Since in fact the red blood cells and the plasma contain almost all of the nonbicarbonate buffer present in extracellular fluid, the concentration of nonbicarbonate buffers in this "total extracellular fluid" is much lower than that of whole blood. However, because there is fairly rapid equilibration between blood and interstitial fluid, from a practical point of view it is totally appropriate to consider these nonbicarbonate buffers as being available to the entire extracellular fluid rather than localized in a subcompartment. The buffering of carbonic acid by the nonbicarbonate buffers of this theoretical "total extracellular fluid" is plotted on the pH-bicarbonate diagram in Figure 4-7 (curve *B–C*). This is the nonbicarbonate buffer curve for "total extracellular fluid," and its slope is far less steep than that of blood because of the much lower concentration of nonbicarbonate buffer. Point *A* is set as the point of reference where HCO_3^- concentration is 24 mmoles/L of fluid, the pH at 7.4, and the pCO_2 40 mm Hg. As the pCO_2 is increased, the system is being titrated with carbonic acid; as shown by curve *A–B*, HCO_3^- increases linearly as the pH decreases. It should again be emphasized that the slope of curve *B–C* in Figure 4-7 is less than curve *B–C* in Figure 4-6 for whole blood in a tonometer because the "total extracellular fluid" has a much lower concentration of nonbicarbonate buffer than does whole blood.

To ascertain if these theoretically derived data just presented are actually true of extracellular fluid under *in vivo* conditions, experiments can be carried out in which either hyperventilation is used to lower the pCO_2 well below the 40 mm Hg of point *A* or the breathing of mixtures of gases containing varying concentrations of CO_2 are used to raise the pCO_2 above 40 mm Hg. Samples of *true* plasma are drawn as a representative source of extracellular fluid and the pH and HCO_3^- concentration of the samples are measured.

At time intervals less than 10 minutes after a *step change** in alveolar pCO_2 the nonbicarbonate buffering power, as indicated by analysis of true plasma, appears to be much greater than would be predicted from curve *B–C* in Figure 4-7, derived from the theoretical analysis of the nonbicarbonate buffer capacity of the "total extracellular fluid." However, this deviation from the predicted values is not due to an error in theory but rather to the fact that the mixing between blood and the interstitial fluid takes time to be complete. Therefore at this early time interval, the true plasma samples are not representative of the total extracellular fluid. When samples of true plasma are taken 10 to 60 minutes after the *step change* in alveolar pCO_2, the data fit almost perfectly the theoretically predicted "total extracellular fluid" nonbicarbonate buffering capacity for carbonic acid (i.e., curve *B–C* in Fig. 4-7); apparently by 10 minutes the mixing of the extracellular fluid is approaching completion. Moreover from 10 to 30 minutes or so, little of the effects of the imposed CO_2 excess or CO_2 deficit in

---

* By *step change* is meant a sudden (almost instantaneous) increase or decrease in the alveolar pCO_2.

the extracellular fluid has been modified by interaction with intracellular buffer systems.

As stated in Chapter 3, it takes about an hour before the intracellular fluid significantly influences the pH changes induced in the extracellular fluid. Therefore for the time interval starting after the mixing of the extracellular fluid is complete and ending when the intracellular fluid has significantly interacted with the extracellular fluid, the pH-HCO_3^- diagram in Figure 4-7 provides good quantitative insight in regard to the capacity of extracellular fluid to buffer carbonic acid. After several hours following a step change in pCO_2 there is again deviation from the theoretical extracellular fluid nonbicarbonate buffering curve for carbonic acid because by this time the extracellular fluid and the intracellular fluid have interacted and compensatory renal activity has occurred to a significant extent.

The intravenous administration of noncarbonic acid (e.g., HCl) to the intact animal can also be quantitatively considered by means of the pH-HCO_3^- diagram. If the pCO_2 is artificially maintained at 40 mm Hg then the buffering of noncarbonic acid can be predicted by use of the pCO_2 40 mm Hg isobar and curve B–C in Figure 4-7; for example, if the pH is dropped to 7.2, 9 mmoles of HCl per liter of extracellular fluid must have been buffered by the bicarbonate buffer system, since the HCO_3^- concentration of true plasma fell from 24 mmoles/L to 15 mmoles/L, point X. Moreover from curve B–C (Fig. 4-7) it is clear that if the pH changed from 7.4 to 7.2 then another 1.7 mmoles of HCl per liter of extracellular fluid must be buffered by the nonbicarbonate buffer system of extracellular fluid in addition to the 9 mmoles buffered by the bicarbonate buffer. From these data it is established that to shift the extracellular fluid pH from 7.4 to 7.2, 10.7 mmoles of HCl must be added per liter of extracellular fluid; it is also shown that 9/10.7 of the HCl would be buffered by the bicarbonate buffer system and 1.7/10.7 by the nonbicarbonate buffer system of the extracellular fluid.

Again practical meaningful data in regard to the extracellular fluid buffering of noncarbonic acid can only be obtained when plasma is sampled after enough time has elapsed to permit the HCl to mix throughout the extracellular fluid space (a half-hour or so) and before much of the H^+ from the HCl has entered the cells (several hours must elapse before much of the H^+ would have entered the cells) or has been excreted by the kidney. It should be noted, however, that a properly functioning ventilatory system (see Chapters 5 and 8) will respond to the increase in $[H^+]$ by increasing alveolar ventilation leading to what is called respiratory compensation, which by lowering the pCO_2 below the 40 mm Hg isobar will shift the pH from 7.2 toward but not to 7.4 and at the same time further decrease the bicarbonate concentration of the extracellular fluid as shown graphically by the arrow in Figure 4-7. Since the decrease in pH due to the HCl will be much reduced because of this respiratory compensation, less of the H^+ of this administered noncarbonic acid is buffered by the nonbicarbonate buffer system when respiratory compensation occurs than when the pCO_2 is held at a pCO_2 of 40 mm Hg.

A similar analysis can be made for intravenous administration of NaOH. If again the pCO_2 is artificially maintained at 40 mm Hg and the extracellular fluid is titrated by the NaOH along the pCO_2 40 mm Hg isobar, the amount of the NaOH buffered by the bicarbonate and non-bicarbonate buffer systems respectively can be readily assessed from the pH-HCO_3^- diagram of Figure 4-7. However, here too a respiratory compensation may occur (although it does not invariably occur) if the pCO_2 is not artificially maintained at 40 mm Hg. Respiratory compensation retains CO_2 and raises the pCO_2 above 40 mm Hg. This retention of CO_2 leads to less of the NaOH being buffered by the nonbicarbonate buffering system than when the pCO_2 is artificially held at 40 mm Hg.

Clearly if the NaOH or the HCl are fully mixed with and remain in the extracellular fluid and if the extent of respiratory compensation is known, the pH-HCO_3^- buffer diagram enables the ready calculation of the amount of noncarbonic acid or alkali being buffered and the distribution of this buffering action between the two types of buffer systems (bicarbonate and nonbicarbonate). However, as length of time increases following the administration of the acid or alkali, use of the extracellular fluid pH-HCO_3^- diagram of Figure 4-7 becomes of little value for a quantitative assessment of the amount of noncarbonic acid excess or alkali excess present. The main reason for this is the interplay between the intracellular fluid and the extracellular fluid compartments which leads after several hours to a significant amount of the added alkali or acid being buffered intracellularly. It should also be noted that the renal system excretes both acid or alkali but that this excretion takes many hours or even days to fully occur. Obviously, when a physician samples the true plasma of a patient who has been suffering from an acid-base problem for many hours or longer he cannot gain from this sample a quantitative assessment of the patients acid-base status but rather only an approximation which provides little or no information on the changes that have occurred in the intracellular fluid compartment of this patient. However, if these data are used in conjunction with the clinical findings and knowledge of previous therapy, they can be a most valuable guide to treatment.

PROBLEMS

1. If the true plasma of arterial blood has a pH of 7.4 when it is equilibrated with pCO_2 36 mm Hg, what is its HCO_3^- concentration?

2. If 1 liter of the blood sample represented by Figure 4-6 is maintained at pCO_2 40 mm Hg and HCl is added to decrease its pH from 7.40 to 7.30, how much HCl must be added to make this change in pH?

3. If true plasma at $38°C$ contains 2 mmoles of phosphate per liter and it has a pCO_2 of 40 mm Hg and a HCO_3^- concentration of 20 mEq/L, what is the concentration of $H_2PO_4^-$ and $HPO_4^=$ in the plasma? The pK' of $H_2PO_4^-$ is 6.8 under these conditions.

REFERENCES

For Discussion of Chemistry of Carbonic Acid:

Edsall, J. T., and Wyman, J.: Biophysical Chemistry, Chapter 10. Academic Press, New York, 1958.

For Discussion of the HCO_3^- Buffer System:

Davenport, H. W.: The ABC of Acid-Base Chemistry. 4th Edition. University of Chicago Press, Chicago, 1958.
Robin, E. D., Bromberg, P. A., and Cross, C. E.: Some aspects of the evaluation of vertebrate acid-base regulation. Yale J. Biol. Med. *41*:448–467 (1969).

For Discussion of Blood Buffers:

Winters, R. W., and Dell, R. B.: Regulation of Acid-Base Equilibrium. *In* Yamamoto, W., and Brobeck, J. (eds.): Physiological Controls and Regulations. W. B. Saunders Co., Philadelphia, 1965.

For Discussion of Buffering in Extracellular Fluid:

Woodbury, J. W.: Regulation of pH. *In* Ruch, T., and Patton, H. (eds.): Physiology and Biophysics. W. B. Saunders Co., Philadelphia, 1965.

5

PROCESSING OF CARBONIC ACID BY THE BODY

SOURCE AND RATE OF CARBONIC ACID PRODUCTION

The processes of energy transformation by which the energy of the chemical configuration of foodstuffs is converted into physiologically useful forms, such as the high energy phosphate bond of ATP, involve a series of metabolic events. Several of these reactions require O_2 as an electron and hydrogen acceptor. Others are decarboxylations which produce CO_2. Most of this CO_2 production occurs from the chemical reactions of the tricarboxylic acid cycle. For every millimole of O_2 consumed in the aerobic metabolism of the mammal, somewhere between 0.7 and 1 mmole of CO_2 is produced by these decarboxylation reactions. From this metabolic activity man produces about 13,000 mmoles of CO_2 per day, or in acid-base terms about 13,000 mmoles of total "carbonic acid" per day.

Usually this large amount of acid production causes no acid-base difficulty because the lungs are able to excrete the CO_2 as rapidly as it is formed metabolically. However, to be excreted the CO_2 must be transported by the blood from the tissues producing it to the lungs. This transport takes place with little change in the blood pH because the nonbicarbonate buffers of blood, particularly hemoglobin, effectively buffer the carbonic acid.

The neural system so controls the rate of alveolar ventilation that a steady-state concentration of dissolved CO_2 of about 1.2 mmoles per liter of arterial plasma and about 1.5 mmoles per liter of intracellular water is normally maintained. This steady-state concentration of CO_2 is compatible with a pH of about 7.4 in the extracellular fluid and one of about 7.0 in the intracellular fluid* (both fluids have nonbicarbonate buffers which function to maintain the pH). It is because the intracellular compartment is where CO_2 is generated by the decarboxylation reactions that the pCO_2 intracellularly must normally be greater than that found in the extracellular fluid. If the elimination of CO_2 by the lungs is impaired and the concentration of CO_2 builds up to levels well above 1.2 mmoles/L of arterial plasma water and 1.5 mmoles/L intracellular fluid water, then an acid-base derangement occurs called respiratory acidosis to be discussed in detail in Chapter 10. Moreover if for any reason alveolar ventilation increases to the point that the concentration of CO_2 in the body fluids falls to levels significantly less than the normal steady-state values, the mammal suffers from a derangement in acid-base physiology known as respiratory alkalosis to be discussed in Chapter 11.

The remainder of this chapter will be focused upon the way in which CO_2 is transported in the mammal from the site of production to the lungs (i.e., the site of elimination) and the way in which the elimination of CO_2 by the lungs is controlled relative to the maintenance of a near constancy of pH in the fluids of the mammal.

ELIMINATION OF CO_2 BY THE LUNGS

The mammalian ventilatory system is able to eliminate the metabolically produced CO_2 at a rate equal to the rate at which it is formed by the tissues and at the same time to maintain the total "carbonic acid pool" at approximately 1.2 mmoles per liter of arterial plasma water. How this is accomplished can be illustrated by considering a normal man at sea level under resting conditions. This man is continually ventilating the alveolar air space with air taken in from his environment; such air is normally almost free of CO_2 ($pCO_2 = 0.3$ mm Hg). The rate at which this environmental air is turned over in the alveoli is called the alveolar ventilation. In normal man at rest (a respiratory state called eupnea) alveolar ventilation is about 4 L/min, and the rate of CO_2 production by the tissues is about 212 ml/min. Clearly if his alveolar ventilation of 4 L/min is to carry away the 212 ml of CO_2 per minute delivered to the alveoli from the tissues by the blood, the concentration of CO_2 in the alveoli must be approximately 5.3 per cent of the alveolar gas mixture. Assuming the barometric pressure is 760 mm Hg, the pCO_2 in the alveoli of this man is approximately 40 mm Hg. Since the arterial blood is in equilibrium with the alveolar air, it will have a CO_2 tension of 40 mm Hg or, more usefully expressed for acid-base

---

* Note that although most investigators find the pH of the intracellular fluid to be ∼7.0, there have been reports of it being as low as 6.0.

purposes, a dissolved CO_2 concentration of about 1.2 mmoles per liter of water.

If for any reason the rate of alveolar ventilation in this man were suddenly increased to 8 L/min with no change in rate of CO_2 production by his tissues, then after a transient period during which some of the pool of dissolved CO_2 would be "washed" out of the body fluids, another steady-state would be reached in which the 212 ml of CO_2 produced per minute would be eliminated by the 8 L/min alveolar ventilation at an alveolar pCO_2 of 20 mm Hg. The new steady-state concentration of CO_2 in the water of arterial blood will be 0.6 mmoles per liter as predicted by Henry's law. Indeed the fall in dissolved CO_2 concentration throughout the body fluids will shift the pH of body fluids to values greater than the normal pH. Such an individual would be said to be suffering from respiratory alkalosis.

On the other hand, if the same normal man reduced his alveolar ventilation suddenly from 4 L/min to 2 L/min, after the transient state during which the concentration of dissolved CO_2 will be increasing in his body fluids, a new steady-state will be reached where the amount of CO_2 produced by the tissues is equal to the amount eliminated via the alveolar ventilation. For this to occur at an alveolar ventilation of 2 L/min the alveolar pCO_2 will have to become 80 mm Hg. Since this pCO_2 is in equilibrium with the arterial blood, the latter will contain about 2.4 mmoles of dissolved CO_2 per liter of water. The increase in dissolved CO_2 in the fluids of this man will cause the pH to be below normal, i.e., this man suffers from respiratory acidosis.

From the foregoing discussion, it is obvious that not only is it important for the lungs of a normal man to eliminate CO_2 from the body as rapidly as it forms but it is equally important that the rate of alveolar ventilation be such that the alveolar pCO_2 is maintained at about 40 mm Hg. Indeed if the metabolic rate of normal man is greatly increased (e.g., during muscular exercise), alveolar ventilation is increased in such a way that the CO_2 is eliminated from the body as rapidly as it is formed and the alveolar pCO_2 remains close to 40 mm Hg.

Obviously man and other mammals have an excellent control system which precisely adjusts alveolar ventilation to conditions as different as rest and vigorous exercise, but the detailed mechanism by which this control system operates is not fully understood. However, the fact that the pCO_2 of the alveolar air can be maintained so close to 40 mm Hg certainly relates at least in part to the presence of chemoreceptors which are sensitive to the concentration of CO_2 and are located in areas of the medulla of the brain bathed by the cerebrospinal fluid. If for any reason the alveolar ventilation of a normal man were to be less than needed to maintain the alveolar pCO_2 at 40 mm Hg, the resulting increase in pCO_2 in the cerebrospinal fluid would stimulate these chemoreceptors which in turn would cause the medullary respiratory centers to increase alveolar ventilation and thus restore the CO_2 content of the body fluids toward normal. By the same token if the alveolar ventilation in a normal man increased so as to reduce the alveolar pCO_2 below 40 mm Hg, then the response of the chemoreceptors in the medulla would cause the medullary respiratory centers to reduce alveolar ventilation and thus restore the CO_2 content of the body fluids toward normal. Therefore, although this control system is not fully understood, enough is known to give a rough idea of the way it regulates the CO_2 concentration of the body fluids. Since this control system is always adjusting the CO_2 concentration, in acid-base terms it adjusts the total "carbonic acid pool" concentration and thus is a fundamental mechanism for the maintenance of a normal pH in the body fluids.

However, there are circumstances where the alveolar ventilation does not function so as to maintain alveolar pCO_2 at 40 mm Hg. For instance, under conditions of high altitude where O_2 is in short supply the falling O_2 tension in the body fluid causes alveolar ventilation to increase. In this case the maintenance of alveolar pCO_2 at 40 mm Hg is sacrificed in the quest for O_2 and the alveolar pCO_2 may fall well below 40 mm Hg, thus leading to a respiratory alkalosis. Another example is seen when an accumulation of noncarbonic acid occurs (i.e., when metabolic acidosis occurs). The fall in pH in the body fluids caused by the accumulation of noncarbonic acid stimulates receptors sensitive to H^+ concentration. They in turn cause alveolar ventilation to increase to levels well above that needed to maintain alveolar pCO_2 at 40 mm Hg. This response has great compensatory value in terms of acid-base regulation since the lowering of the carbonic acid content of body fluids by the increased alveolar ventilation functions to restore body fluid pH toward the normal. In a similar fashion if the pH of body fluids increases due to accumulation of alkali (i.e., metabolic alkalosis), alveolar ventilation is often but not invariably reduced; the resulting increase in carbonic acid levels of the body fluids to some extent mitigates the rise in pH. A detailed discussion of these respiratory compensatory responses to metabolic acidosis and metabolic alkalosis will be presented in Chapters 8 and 9.

ACID-BASE ASPECTS OF CO_2 TRANSPORT BY BLOOD

Although in the normal individual alveolar ventilation maintains the carbonic acid pools of the body fluid at fairly constant levels, there still remains the question of what effect the transport of CO_2 from the tissues to the alveoli has on the pH of the blood. Under normal resting conditions the arterial blood entering the tissue capillaries has a CO_2 tension of 40 mm Hg, and as this blood courses through the capillaries it picks up enough CO_2 to raise the pCO_2 to 46 mm Hg. From the tonometer data on the buffering of CO_2 by blood presented in Chapter 4 (see Fig. 4-6), this increase in pCO_2 should cause the pH of the blood to fall from 7.41 at the arterial end of the capillary to about 7.32 at the venous end. However, in normal

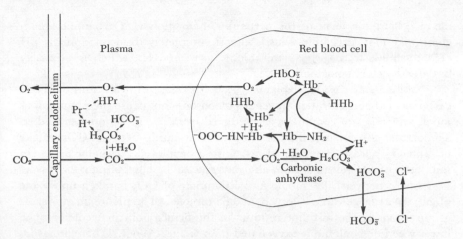

Figure 5-1. Schematic representation of loss of O_2 from blood and uptake of CO_2 by blood in tissue capillaries. Solid lines refer to major pathways and hatched lines to minor pathways. By reversing direction of the arrows, the sequence of events occurring in pulmonary capillaries is depicted.

individuals at rest the pH of the mixed venous blood is 7.38 compared to 7.41 for arterial blood. Obviously under *in vivo* conditions the uptake of carbon dioxide by blood at the tissues causes much less of a fall in the blood pH than predicted from the *in vitro* tonometer data presented in Chapter 4.

The reason for this difference between *in vitro* and *in vivo* data is evident only if the interactions between CO_2 and O_2 transport are considered. In Figure 5-1 a schematic representation of delivery of O_2 to the tissues by blood and the uptake of CO_2 by blood from the tissues is presented. The CO_2 concentration in the tissue cells is higher than it is in the arterial blood entering the tissue capillary; therefore CO_2 diffuses into the blood. It first enters the plasma compartment and then the red blood cell compartment. In the plasma compartment, a small amount of the CO_2 is hydrated to H_2CO_3 which in turn dissociates to HCO_3^- and to H^+ which is effectively buffered primarily by the plasma proteins. The reason for this slow rate of hydration of the CO_2 relates to the absence in plasma of carbonic anhydrase, an enzyme catalyst for this slow reaction.

However, the CO_2 which enters the red blood cells rapidly undergoes two reactions. One is the hydration of CO_2 to form H_2CO_3, catalyzed by carbonic anhydrase which is present in the red blood cell at high concentrations; the H_2CO_3 rapidly dissociates to HCO_3^- and to H^+ which is buffered primarily by the hemoglobin buffer system. Much of the HCO_3^- formed in the red blood cell diffuses into the plasma in exchange for Cl^-. The other reaction that CO_2 undergoes is with the amino groups of the non-oxygenated hemoglobin to form carbamino-hemoglobin. Since the amino groups of the nonoxygenated hemoglobin are far more able to form the carbamino compound than those of the oxygenated hemoglobin, the loss of oxygen from hemoglobin to the tissues facilitates uptake of CO_2 by the red blood cell in the form of the carbamino-hemoglobin. Carbamino-hemo-globin like most carbamic acids is almost completely dissociated at the pH of body fluids to a carboxylate anionic form and H^+ which is primarily buffered by the hemoglobin buffers.

Nothing stated so far explains why the increase in pCO_2 and therefore carbonic acid concentration in venous blood does not cause the pH to fall as much under *in vivo* conditions as predicted from *in vitro* tonometer studies. To consider this further, the quantitative relationship between CO_2 uptake by and O_2 release from the blood must be explored. Usually for every millimole of O_2 released from the oxygenated hemoglobin to the tissues somewhere between 0.7 mmole and 1.0 mmole of CO_2 is taken up by the blood, the average value being about 0.85 mmoles. The pK' of an imidazole group of one of the histidine residues in the hemoglobin molecule is much lower when hemoglobin is oxygenated than when it is not. Therefore as O_2 leaves the hemoglobin molecule and nonoxygenated hemoglobin is formed, this imidazole group removes protons from the aqueous environment; indeed for every millimole of O_2 leaving the hemoglobin molecule this imidazole takes up approximately 0.7 mmoles of H^+ from the medium ,and in the process shifts the medium to a more alkaline pH.

Under *in vivo* conditions, however, as the nonoxygenated hemoglobin is being formed from the oxygenated hemoglobin, CO_2 is simultaneously entering the blood from the tissues and generating H^+. If the extent to which the CO_2 generates H^+ is just sufficient to meet the uptake of H^+ by the non-oxygenated hemoglobin being formed, there will be no change in pH of the blood during the exchange of these two gases, a phenomenon called *iso-hydric buffering*. However under most physiological circumstances, more than 0.7 mmoles of H^+ are formed from CO_2 entering blood per every millimole of O_2 leaving, and the small amount of H^+ above 0.7 mmoles is buffered by the nonbicarbonate buffers of blood, with the usual change in pH seen with such buffering. Therefore, the pH of venous blood is just slightly lower than that of arterial blood unlike what would be predicted from the *in vitro* tonometer data of Figure 4-5 because in the tonometer studies the pO_2 in the gas phase maintains the hemoglobin nearly 100 per cent saturated with O_2 throughout the experiment.

In the alveoli of the lung all of the processes just described go in the reverse direction, with the result that the venous blood delivered to the alveoli by the pulmonary artery increases its pH slightly while coursing through the pulmonary capillaries. Therefore arterial blood has a slightly more alkaline pH than venous blood.

Although not directly related to acid-base regulation, it is probably desirable at this point to quantitatively describe the chemical forms by which CO_2 is transported in the blood. The data for a normal typical man are presented in Table 5-1. Clearly in both the arterial and the venous blood most of the CO_2 is in the form of HCO_3^- of which 75 to 80 per cent is in the plasma. A small but significant amount of CO_2 is present in the blood

Table 5-1. *Values for the Various Chemical Forms of Carbon Dioxide in the Blood of a Typical Man at Rest*

CHEMICAL FORM	ARTERIAL BLOOD (mmoles/L)	MIXED VENOUS BLOOD (mmoles/L)	VENOUS-ARTERIAL DIFFERENCE (mmoles/L)
Dissolved CO_2	1.20	1.34	0.14
HCO_3^-	24.00	25.09	1.09
Carbamino-CO_2	0.97	1.42	0.45

in the form of dissolved CO_2 and in the form of carbamino-CO_2. Some 65 to 70 per cent of the dissolved CO_2 is in the plasma, and almost all of the carbamino-CO_2 is in the red blood cell. The column on the far right in the table indicates the amount of and the form by which the CO_2 entering the blood at the tissues is transported to the alveoli and released there. For every liter of blood going to the alveoli in this resting man about 1.68 mmoles of CO_2 are transported from tissue to alveoli and excreted there. Of this, 1.09 mmoles are in the form of HCO_3^-, most of which was generated within the red blood cell, but over 80 per cent is transported in the plasma. This is so because, as depicted in Figure 5-1, Cl^- of plasma exchanges for the HCO_3^- generated in the red blood cell. Although the carbamino-CO_2 makes up only a small fraction of the total blood CO_2, it is an important transport form because the formation and degradation of this compound is greatly influenced by the oxygenation and deoxygenation of the hemoglobin molecule as previously described. For this reason 0.45 mmoles of CO_2 per liter of blood is transported for excretion in the form of the carbamino-CO_2 compound. A small but significant amount (0.14 mmoles/L) is transported for excretion in the form of dissolved CO_2.

PROBLEMS

1. If a man with a given metabolic rate maintains an alveolar pCO_2 at 40 mm Hg when his alveolar ventilation is 4 L/min, what will his alveolar and arterial blood pCO_2 become if he reduces his alveolar ventilation to 3 L/min without changing his metabolic rate?

2. On the basis of the data presented in Chapter 4, approximately what will the pH of the arterial blood of the man described in Problem 1 be two minutes or so after this decrease in alveolar ventilation from 4 L/min to 3 L/min?

3. On the basis of the data presented in Chapter 4, approximately what will the pH of the arterial plasma of the man described in Problem 1 become 30 minutes or so after this decrease in alveolar ventilation from 4 L/min to 3 L/min?

REFERENCES

For Metabolic Production of CO_2:

Brown, A. C., and Brengelmann, G.: Energy Metabolism. *In* Ruch, T. C., and Patton, H. D. (eds.): Physiology and Biophysics. W. B. Saunders Co., Philadelphia, 1965.

For Regulation of CO_2 *Excretion by Lungs:*

Comroe, J. H., Jr.: Physiology of Respiration. Year Book Medical Publishers, Inc., Chicago, 1965.

For Acid-Base Aspects of CO_2 *Transport:*

Davenport, H. W.: The ABC of Acid-Base Chemistry. 4th Edition. University of Chicago Press, Chicago, 1958.

6

PROCESSING OF NONCARBONIC ACIDS BY THE BODY

SOURCES AND RATE OF PRODUCTION OF NONCARBONIC ACIDS IN NORMAL MAN

The primary source of noncarbonic acid production in normal man is protein catabolism. Protein contains sulfur in chemical forms which do not donate protons to the body water and buffers, specifically the sulfhydryl groups of cysteine residues, the disulfide linkages of cystine residues, and the thioether groups of methionine residues. During the course of the metabolism of proteins this sulfur is converted to $SO_4^=$ with H^+ being simultaneously generated as the cation required for electroneutrality. In other words, the sulfur of the amino acid residues, cysteine, cystine, and methionine is converted to sulfuric acid; e.g., the net chemical reaction for the metabolic conversion of methionine to CO_2, H_2O, urea, and sulfuric acid is:

$$2\,C_5H_{11}NO_2S + 15\,O_2 \rightarrow 4\,H^+ + 2\,SO_4^= + (NH_2)_2CO + 7\,H_2O + 9\,CO_2$$
methionine   urea

A minor source of noncarbonic acid is yielded by the catabolism of phospholipids which contain diesterified phosphate; the diesterified phosphate

$$
\cdots O - \overset{\displaystyle O}{\underset{\displaystyle O^-}{\overset{\displaystyle \|}{P}}} - O \cdots
$$

is converted to $H_2PO_4^- + HPO_4^= + H^+$. There is also the metabolic

69

production of alkali from certain foodstuffs, the details of which will be discussed in Chapter 7. However, whenever protein makes up a large fraction of the diet as it does in most western countries, the amount of metabolic production of noncarbonic acid is greater than the amount of alkali production. The average adult American produces about 40 to 60 mEq of noncarbonic acid in excess of his metabolic production of alkali per day.

Under certain circumstances there are other sources of noncarbonic acid production. For example, during vigorous exercise or under hypoxic conditions lactic acid can be produced in large quantities. In normal individuals lactic acid production is usually a transient phenomenon, i.e., after exercise is completed or the hypoxic condition eliminated, the lactic acid is metabolized to CO_2 and H_2O. In such cases the production of lactic acid results in no more than a transient increase in the noncarbonic acid content of body fluids and one which is eliminated by the metabolic destruction of the acid rather than by renal excretion. The ketone bodies β-hydroxybutyric and acetoacetic acids are noncarbonic acids produced by normal man during fasting. Upon eating, much of these acids that have accumulated are further catabolized to CO_2 and H_2O. Therefore in normal man these acids only transiently cause the body to have an increased load of noncarbonic acid. Of course if for any reason these acids are not further metabolized, they must be processed in a fashion similar to that of sulfuric and phosphoric acids.

Besides the metabolic generation of noncarbonic acid, the diet may contain them as such, i.e., physiologic Brønsted-Lowry acids may be present as such in the diet. For example, acetic acid, i.e., vinegar, is often ingested; however, it only functions transiently as a noncarbonic acid because man is able to rapidly convert it to CO_2 and H_2O. Other examples of such acids are citric, isocitric, and other organic acids present in fruits.

The administration of NH_4Cl leads the metabolic system to form hydrochloric acid as follows:

$$2\ NH_4Cl + CO_2 \rightarrow 2\ H^+ + 2\ Cl^- + H_2O + \underset{\text{urea}}{(NH_2)_2CO}$$

The only way this noncarbonic acid (i.e., HCl) can be eliminated from the body is by renal excretion since the body has no way of metabolizing it further. Of course, aside from being administered experimentally or therapeutically, NH_4Cl is not consumed by man in significant quantities.

BUFFERING OF NONCARBONIC ACIDS IN THE BODY FLUIDS

The sulfuric and phosphoric acids, the major forms of noncarbonic acid produced in normal man, are generated intracellularly during the catabolism of protein and phospholipid. Therefore the intracellular buffers, primarily nonbicarbonate buffers, initially buffer the H^+ of these metabolically generated acids. However, under steady-state conditions the rate of production of H^+ within the cells must be equal to the rate of release of H^+ to the extracellular fluid. Indeed there is excellent evidence that the bicarbonate buffer system of the extracellular fluid, the quantitatively most important buffer system in the extracellular fluid, is involved in buffering the 40 to 60 mEq of excess noncarbonic acid normally produced each day by the catabolism of proteins and phospholipids. In such individuals the HCO_3^- concentration of the plasma is about 24 mEq/L. If this same man is given a diet which gives rise to neither a net production of noncarbonic acid nor a net production of alkali (for instance an appropriate mixture of meat, fruit, and vegetables) the HCO_3^- concentration of his plasma will rise to approximately 27 mEq/L. It is clear therefore that the bicarbonate buffer system of the extracellular fluid had been involved in the buffering of the net noncarbonic acid production of normal man ingesting the usual relatively high protein diet, i.e., the typical diet of man in western countries. Of course all of the other buffer systems of the extracellular fluid are also involved in this buffering, e.g., the hemoglobin buffer system of red blood cells, the phosphate and protein buffer systems of plasma.

When muscle produces lactic acid glycolytically under conditions such as muscular exercise, the lactic acid appears to enter the extracellular fluid rapidly where it is buffered by the bicarbonate and nonbicarbonate buffer systems of the extracellular fluid. Of course this lactic acid is also buffered by the intracellular buffers of muscle as well. It is nevertheless important to note that the production of noncarbonic acid, such as lactic acid by the intracellular enzymatic systems, is handled with remarkable speed and extensiveness by extracellular fluid buffer systems.

Ketone bodies are produced primarily, if not exclusively, by the liver, and there is little doubt that both β-hydroxybutyric and acetoacetic acids are buffered by the hepatic buffer systems. There is also a rapid release of these substances to the extracellular fluid where the buffers of the extracellular fluid, primarily the bicarbonate buffer system, become heavily involved in buffering these noncarbonic acids. With time, interaction between the extracellular fluid and the extrahepatic intracellular buffer systems will occur, and a considerable amount of the H^+ generated by ketone body production in the liver will ultimately be buffered in the intracellular fluid compartment of extrahepatic tissues.

MECHANISMS OF ELIMINATION OF NONCARBONIC ACIDS

The sulfuric and phosphoric acid produced by the catabolism of protein and phospholipid can only be eliminated from the body by the kidney. Indeed any noncarbonic acid that cannot be metabolized to CO_2 and H_2O

must be eliminated by the kidney. Moreover even noncarbonic acids that can be catabolized to CO_2 and H_2O are eliminated by the kidney when present in large excess; e.g., in diabetes mellitus, β-hydroxybutyric and acetoacetic acids are generated much more rapidly than they can be catabolized by the extrahepatic tissues and therefore must be eliminated by the kidney if the H^+ concentration of the body fluid is to remain at levels compatible with life.

Before discussing the way in which the kidney excretes these noncarbonic acids, it is first necessary to describe the renal reabsorption of HCO_3^-. In the normal individual in whom about 40 to 60 mEq of noncarbonic acid are produced from protein and phospholipid catabolism in excess of alkali production, the plasma HCO_3^- concentration is about 24 mEq/L. At this concentration the kidney is filtering approximately 3 mEq of HCO_3^- per minute. The proximal segments of the 3 million or so renal tubules reabsorb about 90 per cent of filtered HCO_3^- returning it to the peritubular capillary blood. The proximal segment tubule cells reabsorb HCO_3^- in quite a circuitous fashion which involves the secretion of H^+ by the proximal tubule cells into the tubular luminal fluid.

Two mechanisms have been proposed for the generation of the H^+ that is secreted into the tubular fluid. One mechanism is described by the following equation:

$$CO_2 + H_2O \xrightarrow{\text{carbonic anhydrase}} H_2CO_3 \longrightarrow H^+ + HCO_3^-$$

The other may be described as follows:

$$\text{redox system} \longrightarrow H^+ + OH^-$$
$$\downarrow$$
$$CO_2 + H_2O \xrightarrow{\text{carbonic anhydrase}} H_2CO_3 + OH^- \longrightarrow HCO_3^- + H_2O$$

Neither mechanism is supported by any really solid experimental evidence. Both mechanisms have in common the following: (1) carbonic anhydrase is required, and (2) for each H^+ generated there is the generation of one HCO_3^-. This relationship between H^+ and HCO_3^- is schematically shown in Figure 6-1. This figure shows further that H^+ is secreted into the tubular lumen in exchange for Na^+. Once in the lumen the H^+ can be accepted by HCO_3^- to generate H_2CO_3. Affixed to the brush borders of the proximal segment tubular cells is a high concentration of carbonic anhydrase (abbreviated as C.A. in the figure) which rapidly converts this H_2CO_3 to CO_2 and H_2O. Most of the CO_2 will diffuse back into the body because H_2O is rapidly reabsorbed in this part of the kidney, and as CO_2 concentration increases it readily diffuses along its concentration gradient.

It should be noted that the Na^+ delivered to the peritubular blood comes from the luminal fluid, but the HCO_3^- delivered to the peritubular blood is generated by the proximal tubule cell. Nevertheless in terms of net

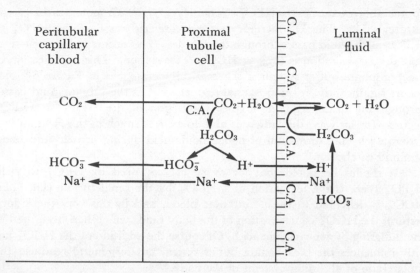

Figure 6-1. Bicarbonate reabsorption by proximal segment. (C.A. is symbol for carbonic anhydrase.)

balance, for every H^+ captured by a HCO_3^- in the tubular luminal fluid, a HCO_3^- has been delivered to the peritubular capillary blood. Therefore in a net sense this circuitous pathway reabsorbs Na^+ and HCO_3^- just as effectively as would a direct transport of both of these ions from the tubular luminal fluid to the peritubular capillary blood. It is established that H^+ secretion by the proximal segment is an active process.

In an individual who generates 40 to 60 mEq of noncarbonic acid per day (i.e., most normal men in western countries) the distal segment of the kidney reabsorbs almost all HCO_3^- not reabsorbed by the proximal segment, i.e., the 10 per cent of the filtered HCO_3^- not reabsorbed by the proximal segment. The mechanism by which the distal segment reabsorbs HCO_3^- is similar to that shown in Figure 6-1 for the proximal segment. However, it does involve secretion of H^+ over much steeper concentration gradients. By this H^+ secretory system the distal segment and the collecting duct can reabsorb all but a very minute amount of the HCO_3^- delivered there.

Of course the kidney does not always reabsorb all of the filtered HCO_3^-. Rather it tends to reabsorb HCO_3^- almost completely until the plasma concentration of HCO_3^- reaches about 27 to 28 mEq/L of plasma, a concentration which might be called the renal threshold for HCO_3^- since at higher concentrations the kidney excretes sizable amounts of HCO_3^-. Of course a normal man, who is excreting 40 to 60 mEq of noncarbonic acid per day, will not have plasma HCO_3^- concentrations as high as 27 to 28 mEq/L and thus his kidney reabsorbs essentially all filtered HCO_3^-.

The H^+ secretion mechanism of the proximal and distal segments not only furnishes H^+ for HCO_3^- reabsorption but also titrates other filtered conjugate bases. The only other conjugate base that normal man filters in appreciable quantities is $HPO_4^=$. The $H_2PO_4^-/HPO_4^=$ molar ratio of the filtrate is $1/4$. The H^+ secretory system converts much of the $HPO_4^=$ to $H_2PO_4^-$ as the fluid passes through the tubules. This occurs at the same time that the H^+ secretion is causing HCO_3^- reabsorption. This acidification of the phosphate buffer system is schematically presented in Figure 6-2 and occurs significantly in the proximal segment but as the tubular fluid passes through the distal segment further conversion of the $HPO_4^=$ to $H_2PO_4^-$ occurs. In this way, the kidney is able to excrete much of the H^+ that was generated by the formation of noncarbonic acid during protein and phospholipid catabolism.

It should be noted that for every H^+ captured by $HPO_4^=$ to yield $H_2PO_4^-$ there is the formation of a HCO_3^- by the renal tubule cell. This HCO_3^- is delivered to the peritubular blood, and in this way the kidney restores the HCO_3^- concentration of the body fluids which had been used in the buffering of noncarbonic acid. Of course the addition of the HCO_3^- not only influences the bicarbonate buffer system but in general readjusts the molar ratio of all buffer systems of the body.

The kidney has still another mechanism of excreting H^+. The renal tubular cells, both the proximal and the distal, are capable of generating NH_3. Since the pH of the tubular cells is about 7.0 most of the NH_3 will immediately capture a H^+ to yield NH_4^+. The following equation depicts the relationship between ammonia and ammonium ion:

$$NH_3 + H^+ \rightleftharpoons NH_4^+$$

Ammonium ion is obviously the acid form and NH_3 the conjugate base; the pK' of this system is about 9.3. Ammonia is very capable of diffusing across cell membranes while NH_4^+ is believed not to diffuse readily.

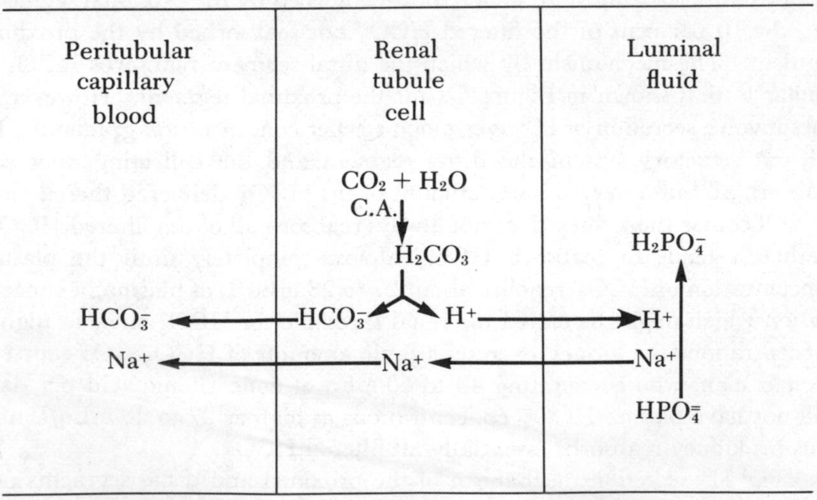

Figure 6-2. Acidification of the phosphate buffer system by renal tubules. (C.A. is symbol for carbonic anhydrase.)

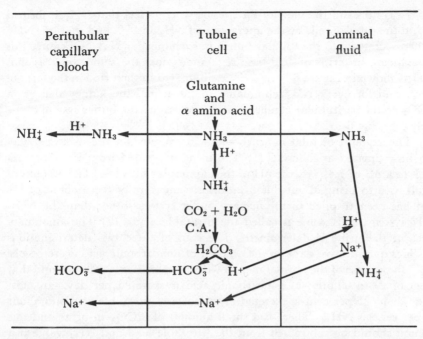

Figure 6-3. A suggested mechanism for the secretion of NH_3 by tubular cells and the excretion of NH_4^+ in urine. (C.A. is symbol for carbonic anhydrase.)

It has been suggested by Pitts that the events depicted in Figure 6-3 occur. Ammonia diffuses from the tubular cells into either the plasma or the tubular luminal fluid, the amount going into each depending on two physical events. One is the relative rate of flow of each of the fluids which is always much greater for plasma flow than for tubular fluid flow. The other factor is the pH of each of these fluids, the lower the pH the more the NH_3 will be trapped as NH_4^+. Since our normal man is excreting acid, his tubular fluid pH falls to 6.0 or lower, and it is this low pH which serves to trap NH_3 into the urine by rapidly converting it to NH_4^+ (Fig. 6-3). Obviously in a condition like metabolic acidosis where the pH of the urine is quite low much more NH_3 will be trapped as NH_4^+ by this pH sink. The important physiological characteristic of this system is that as H^+ is captured by NH_3, H^+ is excreted into the urine as NH_4^+, a substance which does not cause the pH of urine to fall. Since the minimal pH that the urine can reach is 4.4, (i.e., at most the H^+ secreting system of the kidney can cause a 1000-fold H^+ concentration gradient between plasma and urine), it is necessary to capture most of the H^+ secreted into the luminal fluid either by a filtered buffer system such as phosphate or by NH_3 to form NH_4^+. If there were not these systems to capture H^+, the kidney would be able to excrete only very small amounts of noncarbonic acid, e.g., in an unbuffered, NH_4^+-free urine, 1 liter would contain less than 0.1 mEq of H^+. Indeed the amount of H^+ excreted by the kidney in the form of free H^+ is negligible.

In contrast to the view of Pitts, some workers feel that NH_4^+ is formed within the tubular cells by the interaction of NH_3 and H^+ and that this NH_4^+ is then secreted into the tubular lumen in exchange for Na^+. Of course this mechanism, in terms of acid-base regulation, does the same thing as that of Pitts shown in Figure 6-3, i.e., H^+ is eliminated in the urine in the form of NH_4^+, and for every NH_4^+ eliminated in the urine, the kidney delivers a HCO_3^- to the peritubular capillary plasma to restore the buffer system of the body (see Fig. 6-3).

The amount of total noncarbonic acid excreted in the urine per day can be estimated as follows: (1) The urine is titrated from its acidic pH back to a pH of 7.4 (i.e., the pH of the glomerular filtrate). The amount of alkali used to bring about this titration is a measure of the amount of H^+ that has been trapped in the urinary buffer systems other than the NH_3-NH_4^+ system; this value is called the titratable acid. (2) The amount of NH_4^+ in the urine is determined. The sum of these two determinations provides quantitative data on the amount of noncarbonic acid excreted per day. In the case of the normal men in western countries it will be found that a net of 40 to 60 mEq of noncarbonic acid is excreted per day. In most cases about 25 per cent is excreted in the form of titratable acid and about 75 per cent as NH_4^+. There is a small amount of HCO_3^- in urine and this amount should be subtracted from the titratable acid and NH_4^+ value if a totally precise value for the net rate of noncarbonic acid excretion is desired. For every millimole of titratable acid or NH_4^+ excreted in the urine, the kidney has generated and added to the body buffer pools a millimole of HCO_3^-.

Before leaving this area, some general remarks about the elimination of H^+ by the kidney are in order. In normal people most of the H^+ secreted by the kidney tubules is used to reabsorb HCO_3^-. Moreover, normal man does not approach the limiting urinary pH of 4.4. However, in metabolic acidosis the pH of urine may well reach 4.4; in such individuals the rate of H^+ elimination in the urine is not limited by the maximal rate at which H^+ can be secreted by the tubule cells but rather by the limiting concentration gradient for H^+ that the kidney can establish. In such individuals if more buffer or more NH_3 is made available, more H^+ can be excreted in the urine. In a condition such as diabetes mellitus where large amounts of acetoacetate and β-hydroxybutyrate are filtered into the urine, much buffer anion is available to accept secreted H^+. Moreover when acidosis is chronic, the kidney increases its rate of NH_3 production thus making more NH_3 available to capture H^+ from the tubular secreting system. Because of these two mechanisms, a patient with well established diabetic acidosis can excrete very large amounts of noncarbonic acid each day.

Another factor that should be taken into consideration is the effect of the pCO_2 on the H^+ secreting activity of renal tubule cells. The higher the pCO_2 the greater the rate at which the H^+ can be secreted by the tubular cells, and the converse is true when the pCO_2 falls below normal levels.

This relationship between pCO_2 and the rate at which H^+ can be secreted is of great importance in the renal response to respiratory acidosis and respiratory alkalosis (see Chapters 10 and 11). Also the ability to secrete H^+ relates to some extent to the amount of Na^+ available, since for every H^+ secreted a Na^+ must be reabsorbed. In the distal segment and collecting ducts, it is possible that the availability of Na^+ may be a limiting factor in H^+ secretion. Moreover the K^+ concentration in the tubular cell is critical in regard to determining the rate at which H^+ can be secreted. In general, the higher the concentration of K^+ in the tubular cells the less their ability to secrete H^+. It should also be obvious that drugs which function as carbonic anhydrase inhibitors reduce the capacity of the kidney to secrete H^+ and thus to acidify the urine.

PROBLEMS

1. If the minimum pH of the urine of a man is 4.4, approximately what would his daily volume of urine have to be if he had to excrete his daily production of 50 mEq of noncarbonic acid in a urine that contained no buffer system and no NH_4^+?

2. If the pH of the urine is 5.1 and its pCO_2 40 mm Hg and the amount of HCO_3^- filtered by the kidneys is 2.5 mEq/min, the titratable acid in the urine is 100 mEq/day and the NH_4^+ in the urine is 150 mEq/day, and the urine volume is 2 L/day, what is the rate expressed in mEq/day at which the kidneys deliver HCO_3^- to the peritubular blood?

3. The pH of the urine of a subject is 5.1 and the $[CO_2]$ 1.2 mmoles/L and his urine contains 75 mEq of NH_4^+ and 50 mEq of phosphate buffer per liter and no other buffers. If the amount of urine formed each day is 2 liters, at what rate is H^+ being eliminated from the body by means of the renal tubule H^+ secretion system?

REFERENCES

For Sources of and Rate of Production of
Noncarbonic Acids in Normal Man

Camien, N., Simmons, D. H., and Gonick, H. C.: A critical reappraisal of "acid-base" balance. Amer. J. clin. Nutr. *22*:786–793 (1969).

For Buffering of Noncarbonic Acid in the Body
Fluids

Christensen, H. N.: Body Fluids and Their Neutrality. Oxford University Press, New York, 1963.

For Mechanisms of Elimination of Noncarbonic Acids

Ardaillou, R., and Fillastre, J. P.: Proximal tubular acidification in man. Clin. Sci. *37*:655–665 (1969).

Morris, R. C.: Renal tubular acidosis. New Engl. J. Med. *281*:1405–1413 (1969).

Pitts, R. F.: Physiology of the Kidney and Body Fluids. 2nd Edition. Year Book Medical Publishers, Inc., Chicago, 1965.

Pitts, R. F.: The role of ammonia production and excretion in regulation of acid-base balance. New Engl. J. Med., *284*:32–38 (1971).

7

PROCESSING OF ALKALI BY THE BODY

SOURCES OF ALKALI PRODUCTION

Dietary organic anions that can readily be catabolized to CO_2 and H_2O are the major sources of alkali production in man and other mammals. Lactate, citrate, and isocitrate are quantitatively the most abundant of such anions in the diet; all natural foodstuffs contain them, but fruit and vegetables are the most important dietary sources. Usually K^+ and Na^+ are the accompanying cations ingested.

The catabolism of organic anions does not directly lead to a production of OH^- but rather involves the consumption of H^+ as illustrated by Equation 7-1 which depicts the net effect of lactate catabolism:

$$CH_3 - CHOH - COO^- + H^+ + 3\ O_2 \rightarrow 3\ CO_2 + 3\ H_2O \quad (7\text{-}1)$$

As discussed in Chapter 2 the consumption of H^+ is equivalent to the production of alkali. The H^+ consumed by organic anion catabolism cannot be drawn to any significant extent from a free H^+ pool since this is negligible in body fluids (i.e., $\sim 10^{-7}$ M). Rather the various buffer systems of the body donate the H^+ used in Equation 7-1, and in this way the buffers are titrated to a more alkaline pH as shown in Equation 7-2:

$$Na^+ + CH_3 - CHOH - COO^- + HA + 3\ O_2 \rightarrow$$
$$3\ CO_2 + 3\ H_2O + Na^+ + A^- \quad (7\text{-}2)$$

In this case HA and A^- are used to symbolize both the nonbicarbonate and the bicarbonate buffer systems.

79

In most western countries the amount of protein and phospholipid ingested (high meat intake) and catabolized to sulfuric and phosphoric acids is usually much greater than the amount of organic anions ingested primarily from fruits and vegetables, so that a net balance of noncarbonic acid is produced. Therefore people eating such a diet do not have the problem of eliminating a net amount of alkali produced by the catabolism of organic anions. However, in individuals who are primarily vegetarians by choice or who live in countries where fruit and vegetable intake for most of the population far exceeds meat intake, there is net production of alkali each day due to organic anion catabolism. These people must process this net production of alkali (i.e., alkali production in excess of noncarbonic acid production) and eliminate it from the body. The rest of this chapter relates to the mechanisms by which this alkali is processed and eliminated.

BUFFERING OF ALKALI IN THE BODY FLUIDS

Since the catabolism of the organic anions occurs intracellularly, the intracellular buffer systems are the ones which initially donate the H^+ used in the catabolic process illustrated by Equations 7-1 and 7-2. Of course during any sustained period of production of such alkali, a steady-state interaction will be established between the intracellular and the extracellular fluids which involves buffering the alkali by both extracellular fluid buffers and intracellular fluid buffers. This leads in the case of the nonbicarbonate buffer systems to an increase in the more anionic forms, symbolized by Buf^-, relative to the less anionic form, symbolized by HBuf. In the case of the bicarbonate buffer system, this net production of alkali will raise the $[HCO_3^-]$ relative to that of $[CO_2]$; therefore plasma bicarbonate levels rise to levels greater than the 24 mEq/L of the usual normal individual, producing a net amount of noncarbonic acid. Indeed in vegetarians the metabolic production of alkali from the organic anions is great enough to cause the plasma bicarbonate concentration to be greater than 28 mEq/L.

MECHANISMS OF ELIMINATION OF ALKALI FROM THE BODY

The capacity of the kidneys to eliminate alkali in the urine primarily relates to the fact that the $[HCO_3^-]$ in plasma is elevated during net alkali production. This elevated plasma $[HCO_3^-]$ increases the amount of HCO_3^- filtered at the glomeruli to an extent exceeding the rate at which the renal tubules can secrete H^+ at a pCO_2 of $\sim$40 mm Hg (as shown in Fig. 6-1), thus not all filtered HCO_3^- is reabsorbed. In other words, although there is no reduction in the ability of the renal tubules to secrete H^+ in comparison to the normal meat eater, the rate at which H^+ can be secreted is less than that needed to totally reabsorb filtered HCO_3^-. Because of this, significant amounts of HCO_3^- (accompanied by a cation usually Na^+ and K^+) are excreted in the urine. It should be noted that if the pCO_2 is elevated, the capacity to secrete H^+ is markedly increased, and it is for this reason that a person suffering a respiratory acidosis can reabsorb much more HCO_3 than can the normal vegetarian.

Obviously the excretion of significant amounts of HCO_3^- causes the urine to have an alkaline pH (urine has a pCO_2 similar to blood). Under these conditions, buffers filtered by the glomeruli and not reabsorbed will be titrated toward their more anionic forms, e.g., phosphate will be excreted primarily as HPO_4^- rather than as $H_2PO_4^-$ as seen in acidic urine. As a result there will be a net excretion in urine of alkali, primarily as HCO_3^-. It may not be intuitively obvious why the excretion of HCO_3^-, for example, will lead to an increase in the H^+ concentration of the body fluids. This can best be understood by considering the usual dynamics of HCO_3^- relative to the metabolic production and the alveolar elimination of CO_2. Equation 7-3 summarizes this relationship:

In capillaries of the metabolizing tissue:

$$CO_2 + H_2O \rightarrow H_2CO_3 + Buf^- \rightarrow HCO_3^- + HBuf \qquad (7\text{-}3)$$

In alveolar capillaries:

$$HCO_3^- + HBuf \rightarrow Buf^- + H_2CO_3 \rightarrow CO_2 + H_2O$$
$$\downarrow$$
$$\text{expired}$$

Obviously in the capillaries of the tissues producing CO_2, the H^+ generated from CO_2 is accepted by the nonbicarbonate buffer systems ($HBuf:Buf^-$), causing these buffer systems to be titrated to the less anionic form, i.e., the conjugate-acid form. However, in alveolar capillaries H^+ is removed from the nonbicarbonate buffer system and accepted by HCO_3^- to form H_2CO_3 which in turn yields CO_2 and H_2O, the CO_2 being excreted by alveolar ventilation. However, if the kidney excretes HCO_3^-, that which is so excreted cannot be involved in acceptance of a H^+ in the alveolar capillary blood for the generation of CO_2 and H_2O. In a real sense, therefore, the kidney by excreting HCO_3^- in the urine strands a H^+ in the nonbicarbonate buffer systems of the body, i.e., in the form of HBuf. Of course under conditions in which a net production of alkali by the catabolism of organic anions is occurring, the H^+ sequestered in HBuf is used (in a net sense at least) as the source of the H^+ needed for the catabolism of the organic anions. This inter-relationship between the renal excretion of HCO_3^-, the sequestering of H^+ by the nonbicarbonate buffer systems (HBuf), and the use of the H^+ so sequestered by the organic anion catabolizing systems is summarized by Equation 7-4:

$$CO_2 + H_2O \rightarrow H_2CO_3 + Buf^- \rightarrow HBuf + HCO_3^- \rightarrow \text{excreted in urine}$$

$$CH_3 - CH_2OH - COO^- + HBuf + 3\,O_2 \rightarrow 3\,CO_2 + 3\,H_2O + Buf^-$$

$$(7\text{-}4)$$

In summary, when a steady-state is reached in an individual producing a net amount of alkali, e.g., a vegetarian, there is an elevation of plasma [HCO_3^-] and a shifting of the pH of the body fluids to a somewhat alkaline pH relative to that of the usual meat eater who metabolically produces a net amount of noncarbonic acid. This increase in plasma HCO_3^- concentration causes the kidney to excrete significant amounts of HCO_3^- in a urine of alkaline pH. The steady-state relationship that is reached involves the continuous renal excretion of alkali, a phenomenon which makes H^+ available to the buffer systems of the body at a rate equivalent to the rate of urinary alkali excretion. In turn the H^+ made available to the body buffers in this way is continuously used in the catabolism of the organic anions. The steady-state therefore involves renal excretion of alkali, the utilization of H^+ by organic anion catabolism, and a somewhat but not marked elevation in the pH and [HCO_3^-] in the extracellular fluids compared to what is usually considered as normal in western countries.

PROBLEMS

1. If a man catabolizes 200 mmoles of sodium lactate per day and forms 100 mmoles of noncarbonic acid per day, is he producing a net amount of alkali or a net amount of noncarbonic acid? If a net amount of noncarbonic acid or alkali is produced, how much is it in terms of mEq/day?

2. If this man's urine has a pH of 7.8, and his urine contains 50 mmoles of phosphate, how much of this net excretion of noncarbonic acid or of alkali is excreted by the phosphate system?

3. If this man is in a steady-state, i.e., he is excreting his net production of noncarbonic acid or alkali as rapidly as it forms, at approximately what rate is he excreting HCO_3^- (assuming the $HCO_3^-:CO_2$ and $HPO_4^-:H_2PO_4^-$ buffer systems are the only ones present in his urine)?

REFERENCES

For Sources of Alkali Production:

Frisell, W. R.: Acid-Base Chemistry in Medicine. The Macmillan Co., New York, 1968.

For Buffering of Alkali in the Body Fluids:

Christensen, H. N.: Body Fluids and the Acid-Base Balance. W. B. Saunders Co., Philadelphia, 1964.

For Mechanisms of Elimination of Alkalis from the Body:

Pitts, R. F.: Physiology of the Kidney and Body Fluids. Year Book Medical Publishers, Inc., Chicago, 1968.
Wesson, L. G.: Physiology of the Human Kidney. Grune and Stratton, New York, 1969.

8

METABOLIC ACIDOSIS

DEFINITIONS

The following five chapters will deal with the distrubances of acid-base balance commonly encountered clinically. These can be conveniently divided into metabolic and respiratory acidosis and metabolic and respiratory alkalosis. These disturbances may occur simply or in various combinations and are associated with many different clinical disorders. In order to facilitate discussion, certain terminology used in these chapters will now be defined. These disorders will be viewed as abnormal physiological processes of the total organism and not as abnormalities of the blood as an isolated system.

Acidosis is defined as an abnormal condition or process which would produce a fall in the pH or rise in the [H^+] of the blood if there were no secondary changes. Since secondary changes which diminish the extent of pH change do occur it is possible to have an acidosis with a normal blood pH or H^+ concentration.

Alkalosis is defined as an abnormal condition or process that tends to produce a rise in the pH or a fall in the [H^+] of the blood if there are no secondary changes.

The adjective metabolic preceding acidosis or alkalosis means that the primary etiologic factor involves a gain or loss of noncarbonic acid (or fixed acid) or a loss or gain of HCO_3^- by the extracellular fluid. The adjective respiratory preceding acidosis or alkalosis means that the etiologic factor is a decrease or increase in alveolar ventilation with the result that the total "carbonic acid pool" of the extracellular fluid is either too high or too low for the acid-base balance requirements of the organism.

Disturbances of acid-base balance can be "simple" (i.e., be caused by only one primary etiologic factor) or "mixed" (i.e., be caused by more than one primary etiologic factor). For example, a patient may have a "simple" metabolic acidosis, such as diabetic acidosis alone, or a "mixed" metabolic and respiratory acidosis, such as diabetic acidosis and chronic ventilatory insufficiency due to lung disease.

The terms secondary or compensatory may be used to describe a change in the composition of the blood or to describe a process, but such terms should not be used to modify the nouns acidosis or alkalosis.

Acidemia or alkalemia indicates an actual fall or rise in arterial pH respectively. Hypocapnia or hypercapnia refers to a fall or rise in the arterial pCO_2 respectively. Whenever possible, values for pCO_2, pH, or $[H^+]$ should be used rather than these descriptive terms. The following is a tabulation of the normal range of values found for the pH, $[H^+]$, pCO_2, and $[HCO_3^-]$ in arterial blood:

pH = 7.37 to 7.42 $[HCO_3^-]$ = 23 to 25 mmoles/L

$[H^+]$ = 38 to 42 nmoles/L pCO_2 = 38 to 42 mm Hg

The pCO_2 tends to be slightly lower and the pH slightly higher in young females than in males.

By these definitions a metabolic acidosis is an acidosis caused by (1) the increased production, ingestion, or infusion of noncarbonic acid, (2) by the decreased renal excretion of H^+, (3) by an ingress of H^+ from the intracellular to the extracellular fluid and, (4) by a loss of HCO_3^- or other conjugate base from the extracellular fluid which results in an increase in $[H^+]$.

CAUSES OF METABOLIC ACIDOSIS

The causes of metabolic acidosis may be categorized in the manner just mentioned under definitions. Increased production of noncarbonic acid can be caused by diabetic acidosis in which large quantities of the keto-acids, β-hydroxybutyric acid and acetoacetic acid, are accumulated. These acids have low pK' values of 4.39 and 3.58, respectively, and are essentially completely ionized in extracellular fluid. Excess amounts of these keto-acids can also be seen in lesser degrees with starvation ketosis.

Excess production of lactic acid is seen in severe hypoxemia and agonal states as well as in patients receiving large doses of certain drugs, such as Phenformin. It has been reported occasionally in patients severely ill with a variety of diseases ranging from bleeding peptic ulcer to myocardial infarction. It has also been reported to occur spontaneously.

Ingestion of methyl alcohol and salicylates causes acidosis by increasing the accumulation of noncarbonic acids. Paraldehyde and ethylene glycol (anti-freeze) also cause metabolic acidosis but the mechanism involved is not so well defined.

The NH_4^+ contained in ingested NH_4Cl causes acidosis because it is metabolized to urea by the liver with a net yield of one mole of H^+ per mole of metabolized NH_4^+ (see Chapter 6). A similar result occurs when arginine hydrochloride is infused or lysine hydrochloride is ingested.

Loss of HCO_3^- and other conjugate bases as an etiologic factor causing acidosis is seen with severe diarrheas and fistulas causing loss of bowel content. The fluid in the small and large intestine is relatively high in HCO_3^- and in other conjugate bases such as citrate. The range of $[HCO_3^-]$ in small bowel content is about 60 to 100 milliequivalents per liter. With ureterosigmoidostomy, HCO_3^- is lost by exchange across the colonic mucosa into the urine which perfuses this region.

In uremia due to most forms of chronic renal disease, acidosis results because renal ammonia secretion is deficient as is renal H^+ excretion. There is an excess of HCO_3^- in the urine and a decrease in the maximal H^+ gradient between the renal tubular cell and the lumen. In renal tubular acidosis and sometimes in chronic pyelonephritis the major difficulty appears to be an inability to maintain the maximal H^+ gradient between the renal tubular cell and the luminal fluid. The maximal gradient is of the order of 800:1 normally (see Chapter 6) and may be as low as 80:1 in renal tubular acidosis.

BUFFERING MECHANISMS IN THE EXTRACELLULAR FLUID

The most important functional buffer components in blood in the physiologic pH range of 6.8 to 7.8 are hemoglobin, plasma proteins, and bicarbonate. Phosphate is of much less importance because of its low concentration. In the remainder of the extracellular fluid the principal buffering system for noncarbonic acid is the bicarbonate buffer system. Exogenous noncarbonic acids added to the blood are neutralized by all the blood buffers, with the bicarbonate and hemoglobin systems being quantitatively the most important (see Chapter 3).

Since the ratio of HCO_3^- to the total "carbonic acid pool" is approximately 20:1 in extracellular fluid, there is an ample supply of HCO_3^- available for buffering with the added advantage that in the physiological open system the extra carbonic acid formed during the buffering is rapidly dissipated by the lungs as CO_2 as indicated in the following equation:

$$H^+ + HCO_3^- \rightarrow H_2CO_3 \rightarrow CO_2 + H_2O$$

The quantitative aspects of this buffer system have been discussed in Chapter 4.

The bicarbonate buffer system as well as the nonbicarbonate buffer systems are effective almost immediately upon the presentation of an acid load to the extracellular fluid. Thus, a drop in the plasma HCO_3^- concentration occurs relatively rapidly in a person with metabolic acidosis. The other extracellular buffers play a quantitatively less important role in metabolic acidosis than does the bicarbonate buffer system.

COMPENSATORY RESPIRATORY MECHANISMS

The neural system controlling alveolar ventilation is sensitive to the rise in $[H^+]$ which is produced when an acid is added to the extracellular fluid. The mechanism by which an increase in $[H^+]$ stimulates alveolar ventilation is not fully understood but appears to involve receptors in the central nervous system and the peripheral chemoreceptors in the aortic and carotid bodies. The resultant increase in alveolar ventilation is accomplished by an increase in both the tidal volume and the frequency of breathing. The increase in tidal volume is somewhat greater than the increase in the rate of breathing.

As a result of this increase in alveolar ventilation the excess CO_2 produced from the buffering of the added H^+ by the bicarbonate buffer system is excreted. Moreover, if the individual has normal respiratory centers and a normal ventilatory apparatus the arterial pCO_2 is actually reduced to levels below the normal range of 38 to 42 mm Hg. This reduction of arterial pCO_2 will decrease the $[H^+]$ in the blood according to the following equation:

$$[H^+] \text{ nmoles/L} = \frac{23.8 \times pCO_2 \text{ (mm Hg)}^*}{[HCO_3^-] \text{ mmoles/L}} \qquad (8\text{-}1)$$

Or if we wish to express the data as pH rather than H^+ concentration:

$$pH = 6.1 + \log\left(\frac{[HCO_3^-](\text{mmoles/L})}{0.03 \times pCO_2 \text{ (mm Hg)}}\right) \qquad (8\text{-}2)$$

Except in patients with damaged respiratory centers or those with diseases causing impairment of the ventilatory apparatus, the respiratory response to metabolic acidosis is extremely prompt and predictable, occurring within a matter of minutes after the onset of metabolic acidosis. This response can be used clinically to assess the severity of the acidosis in patients with normal ventilatory and neural systems. Failure to achieve the expected degree of response implies a disordered ventilatory or neural system. In order to interpret data regarding the ventilatory response one must know the inter-relationships between the degree of reduction of $[HCO_3^-]$ in arterial blood, the compensatory reduction in arterial pCO_2, and the resultant change in $[H^+]$ of the blood.

---

* The value 23.8 is obtained by multiplying 0.0301 (the solubility coefficient of CO_2 in plasma) by 794. The value 794 is the apparent dissociation constant, K', for the carbonic acid-bicarbonate buffer system when $[H^+]$ is expressed in nanomoles per liter rather than the more conventional moles per liter.

BUFFERING MECHANISMS OF INTRACELLULAR FLUID

The intracellular fluid is not a homogeneous compartment, and the net flux of Na^+, K^+ and H^+ between bone and extracellular fluid, skeletal muscle and extracellular fluid, and cardiac muscle or any other tissue and extracellular fluid is not identical. Nevertheless, the generalization can be made that an increase in extracellular H^+ concentration results in a readjustment of the fluxes of H^+, K^+, and Na^+ so that the intracellular $[H^+]$ increases and the $[K^+]$ and $[Na^+]$ decrease. The rate of diffusion of H^+ into cells is relatively slow compared to that of CO_2. In acute states, in which there is an excess of noncarbonic acid in the extracellular fluid, H^+ is very rapidly buffered by the extracellular fluid buffers but a long time is required before the intracellular buffers become involved. However, ultimately more noncarbonic acid is buffered intracellularly than extracellularly. The movement of H^+ into the cells is reflected in the extracellular fluid by an increase in $[K^+]$.

RENAL MECHANISMS FOR PROCESSING EXCESS ACID

The kidneys are able to increase their rate of excretion of H^+ up to 10 times that of their normal rate, i.e., to about 500 milliequivalents of H^+ per 24 hours, in the presence of a severe metabolic acidosis. The minimum pH of urine is about 4.4. However, the renal elimination of a large noncarbonic acid load is relatively slow, taking place over a period of hours or days; this should be compared with extracellular buffering which occurs in seconds and with respiratory compensation which is measured in minutes.

During metabolic acidosis due to extra-renal causes, virtually all of the HCO_3^- filtered at the glomerulus is reabsorbed by the renal tubules. The limiting factor in renal tubular H^+ secretion is the attainable lumen-peritubular H^+ gradient. As a result, net H^+ excretion is increased but total H^+ secretion is decreased because the lowered renal filtered load of HCO_3^- requires less H^+ secretion for its reabsorption. During experimental chronic metabolic acidosis the renal excretion of ammonium (NH_4^+) increases, even though the urinary pH is not reduced below that seen in acute metabolic acidosis. Apparently the chronic acidosis *per se* causes an increased tubular secretion of NH_3 into the urine as a result of a metabolic adaptation of the renal tubular cells. For further discussion of the mechanism of renal excretion of H^+ see Chapter 6.

VARIOUS COURSES OF ACUTE AND CHRONIC METABOLIC ACIDOSIS

The responses to metabolic acidosis discussed in the preceding sections of this chapter occur in varying magnitude and over different time courses.

For example, the renal response is slow and prolonged with excretion of only a moderate quantity of the excess H^+. However, it will remain effective over long periods of time such as weeks or months. The respiratory response is very prompt and initially of great magnitude, but as metabolic acidosis persists, the ventilatory compensation becomes less effective, possibly because the muscular effort needed for the maximal alveolar ventilation of acute metabolic acidosis can not be sustained over long periods of time. In addition, when blood pH drops below about 7.10 (H^+ concentration greater than 80 nmoles per liter) respiratory minute volume does not increase further but may in fact begin to decrease.

It seems clear that there is a limit to the ventilatory response to metabolic acidosis, and from data obtained in clinical laboratories dealing with acid-base parameters many consider a pCO_2 of about 12 mm Hg to be the limit to which respiratory compensation can be carried. This amount of compensation occurs with a blood H^+ concentration of about 80 nmoles per liter in an uncomplicated metabolic acidosis. However, in our laboratory pCO_2 values as low as 9 mm Hg have occasionally been seen in what we felt to be cases of uncomplicated metabolic acidosis.

The relationship between $[H^+]$, pCO_2, and $[HCO_3^-]$, seen in the intact human being in uncomplicated metabolic acidosis, can be expressed in terms of the range of response in one parameter, such as pCO_2, which can be expected for a given H^+ increment or HCO_3^- decrement due to metabolic acidosis.

From statistical analysis of data from selected patients collected by Albert, Dell, and Winters it can be concluded that in an uncomplicated metabolic acidosis of sufficient severity to increase the blood H^+ concentration to 70 nmoles/L (a pH of 7.14) the ventilatory response should reduce the arterial pCO_2 to about 18 mm Hg, with a range of variation from about 12 to 22 mm Hg. As can be seen from Figures 8-1, 8-2, and 8-3 the relationship between pCO_2 and HCO_3^- in these patients has been found to have less scatter than that between pCO_2 and H^+ or pH. The reasons for this are twofold:

(1) Some correlation between pCO_2 and HCO_3^- is to be expected even in the absence of a physiological mechanism because of the physiochemical relationship between these two parameters as expressed in Equation 8-3 which is a rearrangement of Equation 8-1:

$$[HCO_3^-] = \frac{pCO_2 \times 23.8}{[H^+]} \tag{8-3}$$

Under the circumstances of this study the $[H^+]$ varied only within the narrow limits of 50 to 80 nmoles/L. This relatively small magnitude of change in the denominator of the right side of Equation 8-3 assures a good correlation between pCO_2 and HCO_3^-. (2) Also, since pH or $[H^+]$ and pCO_2 are the independently determined variables and HCO_3^- is derived by

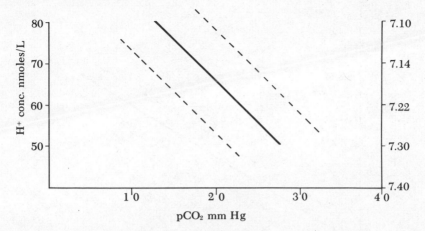

Figure 8-1. Relationship between pCO_2 and $[H^+]$ of arterial blood in metabolic acidosis.
Regression equation: $pCO_2 = 52.95 - 0.50 [H^+]$
S. E. = 2.76 mm Hg
$r = 0.83$
(Modified from Albert, M., Dell, R., and Winters, R. : Ann. Int. Med, *66*:312, 1967.)

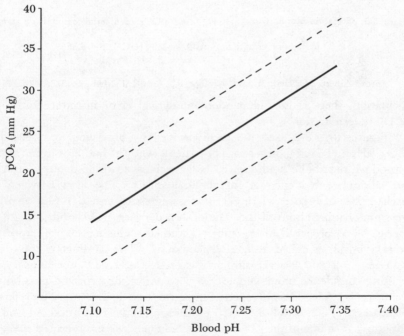

Figure 8-2. Relationship between pH and pCO_2 of arterial blood in metabolic acidosis.
Regression equation: $pCO_2 = -484.64 + 70.26$ pH
S. E. = ± 2.72 mm Hg
$r = 0.83$
(Modified from Albert, M., Dell, R., and Winters, R.: Ann. Int. Med., *66*:312, 1967.)

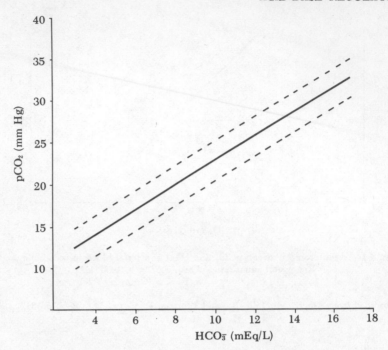

Figure 8-3. Relationship between [HCO_3^-] and pCO_2 of arterial blood in metabolic acidosis.

Regression equation: $pCO_2 = 1.54\,[HCO_3^-] + 8.36$
S. E. $= \pm 1.11$ mm Hg
$r = 0.97$

(From Albert, M., Dell, R. and Winters, R.: Ann. Int. Med., *66*:312, 1967.)

calculation, there is no independent analytical error introduced into the HCO_3^- determination.

Figures 8-1, 8-2, and 8-3 were derived by observing and recording pCO_2, pH, and HCO_3^- in a series of patients who (a) had a disorder known to produce metabolic acidosis, (b) had received no prior therapy, and (c) had laboratory evidence of metabolic acidosis. The area between the hatched lines in Figure 8-1 represents the regression equation plus or minus two standard deviations of the relationship between pCO_2 and [H^+] of the arterial blood observed in the subjects studied. This method of expressing the relationship is called a "confidence band." The parameters considered in Figures 8-2 and 8-3 are similarly expressed as confidence bands. The resulting confidence bands include 95 per cent of the expected possibilities. The use of confidence bands such as these can best be illustrated by referring to Figures 8-1, 8-2, and 8-3 to help us interpret some data obtained from an individual with one of the causes of metabolic acidosis enumerated earlier in this chapter.

Let us say a patient presented with the following values in the arterial blood: pH $= 7.20$, [H^+] $= 64$ nmoles/L, $pCO_2 = 20$ mm Hg, [HCO_3^-] $= 7.5$ mmoles/L. Inspection of Figure 8-2 would confirm that a H^+ concentration of 64 nmoles/L due to metabolic acidosis should result in an increased alveolar ventilation, sometimes referred to as "respiratory compensation," to the extent of a reduction in arterial pCO_2 to a range of 15 to 27 mm Hg. The value obtained in this patient was 20 mm Hg which lies well in the center of that range. The same result would be obtained if pCO_2 and $[H^+]$ or $[HCO_3^-]$ and pCO_2 were considered in Figures 8-1 and 8-3.

The only thing this correlation means is that these values are compatible with the relationship between $[H^+]$ (or pH), $[HCO_3^-]$, and pCO_2 values usually seen in uncomplicated metabolic acidosis. Because these values lie between the limits seen in 95 per cent of cases of uncomplicated metabolic acidosis does not unequivocally establish that this patient suffers from an uncomplicated metabolic acidosis. These same values can be found in patients with various combinations of mixed acid-base disturbances. As will be seen later, a knowledge of the clinical course, prior therapy, current therapy, and other blood chemistry, such as the BUN (blood urea nitrogen), Cl^-, Na^+, and K^+ are necessary to firmly establish whether this is a mixed disturbance or a simple metabolic acidosis.

This does not negate the value of the confidence band, but merely serves to put it in the proper perspective. When viewed from another aspect the confidence band can be even more valuable. If we consider a patient with the same pH of 7.20, $[H^+]$ of 64 nmoles/L, and a pCO_2 of 35 mm Hg it is evident from Figure 8-1 that the pCO_2 lies well above the expected upper limit of 27 mm Hg, and thus we are immediately made aware that this is not a case of simple metabolic acidosis but one complicated by some other factor, most likely a respiratory acidosis. Knowledge that this is a mixed disturbance leads to a different therapeutic approach. Some would refer to this situation as a metabolic acidosis with inadequate alveolar ventilation or metabolic acidosis without adequate respiratory compensation. We have referred to this mixed disturbance as a metabolic acidosis with a concomitant respiratory acidosis. The difference is largely one of semantics, but it is felt that the latter terminology is more likely to call attention to the fact that therapy may be indicated for a ventilatory problem.

Physiologists and biochemists have made much use of the pH-HCO_3^- diagram (described in Chapter 4) to consider metabolic acidosis and other acid-base disturbances. Some of the terminology developed with the use of this diagram is in apparent conflict with that just described, but the differences are actually only semantic. It is probably of value now to discuss first metabolic acidosis in the classic terms used with the pH-HCO_3^- diagram and then to resolve briefly certain of the semantic differences with the newer terminology primarily used in this text.

The pH, $[HCO_3^-]$, and pCO_2 of true arterial plasma of a patient can be represented as a point on the pH-HCO_3^- diagram shown in Figure 8-4. If that point falls at or near point A the patient is probably in a normal acid-base state, but if the point is well removed from point A and falls within area I, the patient probably suffers from a metabolic acidosis. If the point from a metabolic acidosis patient falls close to the hatched line A–E (a line representing the maximal ventilatory effort of a normal person in response

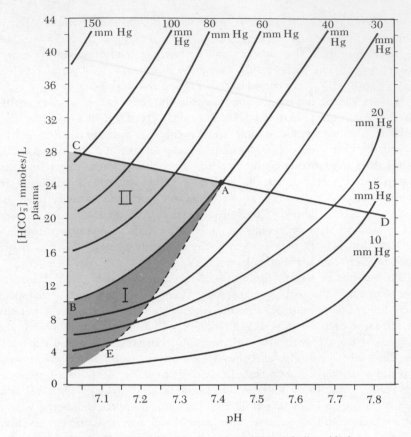

Figure 8-4. pH-HCO$_3^-$ diagram depicting metabolic acidosis.

to the stimulus of extracellular fluid [H$^+$] increase), then the patient suffers from metabolic acidosis accompanied by good ventilatory compensation. If on the other hand the point falls in area I near the *A–B* segment of the pCO$_2$ 40 mm Hg isobar, then the patient is said to be suffering from metabolic acidosis with little or no respiratory compensation. Obviously from this analysis whenever the pCO$_2$, [HCO$_3^-$], and pH of true plasma leads to a point plotted in area I which is sufficiently removed from point *A*, the patient is said to suffer from a simple metabolic acidosis, and the closer the point falls to the *A–E* line the greater the ventilatory compensation present. However, if the point plots in area II, i.e., somewhere between the pCO$_2$ 40 mm Hg isobar segment *A–B* and the *A–C* segment of the nonbicarbonate buffer line, then the patient is said to be probably suffering from a mixed disturbance of metabolic acidosis plus respiratory acidosis. The fact that the pCO$_2$ is elevated and the pH lowered signals a respiratory acidosis (see Chapter 10) and the fact that the [HCO$_3^-$] is less than predicted by the nonbicarbonate buffer line for a given pH and pCO$_2$ (and the [H$^+$] is above normal) indicates the presence of a metabolic acidosis component.

The discussion in most of this text has differed from this classic analysis of metabolic acidosis in only one regard; namely, the term simple metabolic acidosis has been restricted to that area of the graph within the confidence band the center of which is expressed by the curve A–E. Thus by this newer terminology all patients with true plasma $[HCO_3^-]$, pCO_2, and pH which plot outside of the confidence band of line A–E but plot somewhere within either area I or area II are designated as mixed disturbances of metabolic acidosis plus respiratory acidosis. Such a description not only better fits the recently generally accepted definitions of acidosis and alkalosis but also has greater therapeutic usefulness.

Of course the plotting of a single point in time on the pH-HCO_3^- diagram does not provide information in regard to the course of the disease. One must also know the time course of the development of the acid-base disturbance or disturbances and in this regard the clinical history is of prime importance. Also one must consider whether the patient is in a steady-state with renal excretion of acid equal to its production or in a transient state of increasing or decreasing intensity of acid-base dysfunction. Serial determinations over a period of time with the values plotted on the diagram will provide this information.

REFERENCES

For Definitions:

Statement on Acid-Base Terminology. Report of the ad hoc Committee of the New York Academy of Sciences Conference, November 23–24, 1964. Ann. intern. Med. *63*:885–890 (1965).
Winters, R. W.: Terminology of acid-base disorders. Ann. intern. Med. *63*:873–884 (1965).

For Causes:

Doe, R. P.: Metabolic acidosis—nondiabetic. Arch. intern. Med. *116*:717–728 (1965).
Waters, W. C., Hall, J. D., and Schwartz, W. B.: Spontaneous lactic acidosis. Amer. J. Med. *35*:781–793 (1963).

For Renal Mechanisms for Processing Excess Acids:

Pitts, R. F.: Physiology of the Kidney and Body Fluids. Year Book Medical Publishers, Inc., Chicago, 1968.
Steinmetz, P. R.: Excretion of acid by the kidney—functional organization and cellular aspects of acidification. New Eng. J. Med. *278*:1102–1108 (1968).

For Various Courses of Acute and Chronic Metabolic Acidosis:

Albert, M. S., Dell, R. B., and Winters, R. W.: Quantitative displacement of acid-base equilibrium in metabolic acidosis. Ann. intern. Med. *66*:312–322 (1967).
Lennon, E. J., and Lemann, J.: Defense of hydrogen ion concentration in chronic metabolic acidosis. Ann. intern. Med. *65*:265–274 (1966).

CHAPTER

9

METABOLIC ALKALOSIS

DEFINITIONS

Alkalosis as defined in Chapter 8 is an abnormal condition or process that tends to produce a rise in the pH of the blood or a fall in its H^+ concentration if there are no secondary changes.

Metabolic alkalosis is an alkalosis in which the primary cause is a decrease in the H^+ concentration of the extracellular fluid due to either a decrease in noncarbonic acid or an increase in alkali. Theoretically this can result from a decrease in the production of noncarbonic acid, from a loss of noncarbonic acid from the body, from a shift of H^+ from the extracellular to the intracellular space, or from an excess of bicarbonate (HCO_3^-) or other conjugate base which in turn causes a net decrease in the H^+ activity of the extracellular fluid.

CAUSES OF METABOLIC ALKALOSIS

The decreased metabolic production of noncarbonic acid does not cause a clinically significant metabolic alkalosis; however, excessive loss of noncarbonic acid from the body is a fairly common cause of metabolic alkalosis. The loss of gastric juice which contains HCl can lead to metabolic alkalosis in conditions such as severe or prolonged vomiting, in certain types of external fistulas, and in patients receiving gastric suction. In these conditions there is depletion of extracellular volume which contributes to the metabolic alkalosis.

When too much Na^+ and HCO_3^- are retained by the kidney, as happens during excessive use of adrenal steroid hormones or in diseases where these

94

hormones are produced in excess, an alkalosis may result. When alkalosis involves a total extracellular fluid and electrolyte loss, the deficit in total extracellular volume and Cl^- plays a major role in sustaining the alkalosis as will be discussed later.

Disturbances in potassium metabolism resulting in K^+ depletion can cause metabolic alkalosis. K^+ is poorly conserved by the kidney, and losses will continue in the urine even in the absence of intake of this ion. Dietary K^+ depletion will occur only rarely if a person is allowed free selection of food but it can easily occur if a patient is being fed by nasogastric tube or intravenously for a prolonged period of time and K^+ is not added to the administered solutions. K^+ losses can also occur with diarrhea or chronic use of laxatives because the content of the lower bowel is fairly rich in this ion. With chronic renal insufficiency there may be K^+ loss if there is excessive vomiting and diarrhea. Also excessive therapeutic use of cation exchange resins will result in a K^+ deficiency. Moreover in the polyuric phase of recovery from acute renal insufficiency large quantities of K^+ can be lost in the urine. Urinary losses of K^+ are enhanced by most diuretics, including ethacrynic acid, furosemide, and the thiazide diuretics. An increase in adrenal steroid hormones, as seen with Cushing's syndrome or primary aldosteronism and occasionally secondary to certain extra-adrenal tumors such as bronchogenic carcinoma, will cause excessive losses of K^+ in the urine. At present the most frequent cause of adrenal hormone excess is the use of these hormones and their analogues in therapeutic situations.

During K^+ depletion, most K^+ loss occurs from the intracellular fluid where it is the major cation. As intracellular K^+ is depleted, the fluxes of Na^+ and H^+ across all membranes are altered so that a net entry of Na^+ and H^+ into the intracellular space occurs.

If there exists a stimulus for renal Na^+ reabsorption such as an extra-cellular volume depletion when a K^+ deficit is also present, the result will be an inappropriately great excretion of H^+ in the urine. The basis for this is that the filtered, poorly reabsorbable anions, such as phosphate, sulfate, and organic acids, confront the kidney with the necessity of excreting an equivalent amount of cation—either Na^+, K^+, or H^+. Because of extracellular volume depletion, Na^+ is avidly reabsorbed and thus is not available for excretion. If the labile cellular stores of K^+ are also depleted then the demand for cation excretion is met by H^+ excretion, resulting in a metabolic alkalosis.

Metabolic alkalosis may also be produced by an excess of conjugate bases, such as sodium bicarbonate, either because they are ingested or infused or because the kidney excessively reabsorbs them. Also the rapid contraction of the extracellular fluid space following an acute diuresis can cause a temporary relative increase in extracellular $[HCO_3^-]$ termed a "contraction alkalosis." Other conjugate bases, such as that of tris-hydroxy-methylamine methane (THAM), may be used in certain therapeutic situations, but their overuse can also result in metabolic alkalosis. This type of acute metabolic alkalosis is rapidly handled by the kidney through an enhanced urinary HCO_3^- excretion and thus the alkalosis will not persist unless the offending agent continues to be administered in large quantities or unless there are other factors modifying the ability of the kidney to excrete HCO_3^- such as Na^+, K^+, Cl^-, or volume depletion. For a detailed discussion of renal HCO_3^- excretion see Chapter 7.

In a clinical setting metabolic alkalosis is usually caused by multiple physiologic derangements. One of the most common causes of metabolic alkalosis is the use and overuse of various diuretics which cause metabolic alkalosis in a variety of ways. Two common effects of the diuretics most used clinically, i.e., mercurials, furosemide, ethacrynic acid, and the thiazides, are depletion of Cl^- and extracellular volume depletion. The decrease in extracellular fluid volume results in a continued stimulus for Na^+ retention by the kidney. In the face of this stimulus there exists a deficit of the major reabsorbable anion, i.e., chloride, so that HCO_3^- is fully reabsorbed and cation exchange is accelerated with resultant increased K^+ excretion and eventual K^+ depletion and increased H^+ excretion in the urine. The pattern of urinary excretion of these ions varies with different diuretics. For example, ethacrynic acid produces a marked increase of net H^+ excretion in the urine as titratable acid and NH_4^+ with little change in HCO_3^- excretion. Chlorothiazide does not increase urinary H^+ excretion but causes a marked rise in K^+ excretion in the urine which leads to the replacement of intracellular K^+ by H^+. The acid-base effect of all the diuretics is similar, namely, a metabolic alkalosis with Cl^- and extracellular volume depletion.

The challenge presented to the kidney by the metabolic alkalosis of gastric juice loss is similar to that stemming from overly enthusiastic diuretic therapy. It presents the kidney with the choice of either rejecting filtered Na^+ because of a lack of reabsorbable anion (i.e., Cl^-) or reabsorbing the Na^+ by accelerated Na^+-cation exchange with resultant loss of H^+ and K^+ in the urine. The maintenance of volume usually takes precedence in this type of situation over the maintenance of acid-base and K^+ balance.

Metabolic alkalosis is also a frequent complication of therapies used in patients with chronic pulmonary insufficiency. The factors playing a role in the genesis of alkalosis in these cases are low-salt diets, diuretics, and the therapeutic use of adrenal steroids. Even in the absence of these factors, patients with chronic pulmonary insufficiency and CO_2 retention usually have a rise in renal acid excretion and a chloruresis with a resultant increase in extracellular $[HCO_3^-]$. If arterial CO_2 tension is rapidly lowered to normal by treatment with a ventilator, restoration of the normal acid-base relationship requires a bicarbonate diuresis, but this cannot be achieved if there is not enough Cl^- available from the diet to be reabsorbed instead of HCO_3^-. If Cl^- is insufficient, a HCO_3^- diuresis can only be achieved by increased excretion of Na^+ or K^+, but in this situation strong stimuli exist for Na^+ and K^+ conservation and little HCO_3^- diuresis can be expected. As a result of this set of circumstances a high rate of tubular exchange of Na^+ for

H^+ will persist and extracellular HCO_3^- will remain elevated unless sufficient Cl^- is supplied.

The mechanism for the development of metabolic alkalosis in response to excessive adrenal steroid hormones is not clear. The explanation that Na^+ and H^+ exchange in the distal tubule of the kidney is accelerated thus causing HCO_3^- to be generated and retained and H^+ lost in the urine does not fully explain the observed facts. Na^+ retention and K^+ depletion appear to play a role in the development of the metabolic alkalosis, and dietary deficiency of K^+ in the presence of adequate Na^+ seems to accelerate its development. The metabolic alkalosis often seen in patients with adrenal tumors that secrete aldosterone is present in only two-thirds to three-fourths of the cases, and is not seen when secondary aldosteronism develops in diseases such as cirrhosis or malignant hypertension where secretion rates of aldosterone can be equal to or higher than with primary aldosteronism.

In summary then, clinical conditions causing metabolic alkalosis are (1) loss of gastric juice; (2) diuretic therapy with the thiazides, ethacrynic acid, furosemide, and mercurial diuretics; (3) adrenal steroid therapy and, less commonly, endogenous steroid excess, such as Cushing's syndrome, and primary aldosteronism; (4) overly enthusiastic therapy with HCO_3^- or lactate; (5) occasionally excessive oral alkali intake in the form of HCO_3^-; (6) any condition causing persistent K^+ depletion; and (7) after ventilator therapy in a patient suffering from chronic respiratory acidosis. In addition, there is a metabolic and respiratory alkalosis commonly seen in hepatic failure usually secondary to cirrhosis of the liver.

SIGNS AND SYMPTOMS

The signs and symptoms associated with metabolic alkalosis are not distinctive for that condition alone and can occur in many clinical situations not involving metabolic alkalosis. For example many of these symptoms can be produced by hypokalemia alone. The commonest symptoms are nausea, vomiting, mental confusion, and unreliability, progressing at times to drowsiness and coma. Tetany may be observed. The signs of K^+ deficiency (hypokalemia) include weakness, decreased tendon reflexes, paralytic ileus, and abdominal distention as well as tachycardia, various cardiac arrythmias, and increased susceptibility to digitalis intoxication.

INTRACELLULAR AND EXTRACELLULAR BUFFERING MECHANISMS

If alkalosis is induced by HCO_3^- infusion in a nephrectomized dog, about $\frac{2}{3}$ of the added HCO_3^- remains in the extracellular compartment and

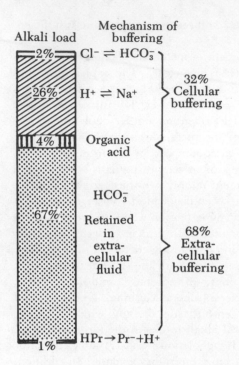

Figure 9-1. Mechanisms of buffering of HCO_3^- infused intravenously in nephrectomized dogs. (From Pitts, R. F.: Physiology of the Kidney and Body Fluids, Year Book Medical Publishers, Inc. Chicago, 1968.)

is buffered there while $\frac{1}{3}$ is buffered in a circuitous and not totally understood way by the intracellular system. This is illustrated graphically by Figure 9-1. In addition, lactic acid generated in the intracellular space is released to the extracellular fluid causing conversion of HCO_3^- and other conjugate bases to their conjugate acids. If the kidneys were intact and there were no other electrolyte abnormalities, the alkalosis would be rapidly corrected by augmented renal HCO_3^- excretion.

COMPENSATORY RESPIRATORY MECHANISMS

While the respiratory response to metabolic acidosis in the presence of a normal ventilatory apparatus and central nervous system is prompt and predictable, the ventilatory response to metabolic alkalosis is quite variable. There are conflicting studies in the literature but recent work seems to show that metabolic alkalosis induced by sodium bicarbonate infusion, THAM, and ethacrynic acid produce a compensatory reduction in alveolar ventilation. This is accomplished primarily by decreasing the resting tidal volume, the frequency of breathing remaining relatively unchanged. A decrease in alveolar ventilation is not observed when similar degrees of extracellular metabolic alkalosis are produced by thiazide diuretics and aldosterone.

The agents which result in decreased alveolar ventilation are buffers or agents which increase urinary H^+ excretion; moreover they produce only a slight to moderate K^+ loss in the urine. Those agents which do not result in decreased alveolar ventilation produce little H^+ loss in the urine but fairly large urinary K^+ losses. The ventilatory response to inhalation of a gas mixture containing 5% CO_2 is impaired in those subjects who have decreased alveolar ventilation accompanying their metabolic alkalosis. The response to CO_2 inhalation is not impaired in those who do not compensate for metabolic alkalosis by elevating their arterial pCO_2 through reduced alveolar ventilation.

DETERMINING FACTORS IN RENAL RESPONSE TO METABOLIC ALKALOSIS AND THE RELATIONSHIPS BETWEEN pH, pCO_2, AND [HCO_3^-]

Any excess of conjugate base added to the extracellular fluid will rapidly lead to a relatively large increase in the HCO_3^- concentration of the extracellular fluid. This in turn will cause a large increase in the HCO_3^- load presented to the renal tubule by the glomerular filtrate. As an example, an increase of the plasma HCO_3^- level from 23 to 31 mmoles/L would result in an increase of 1440 mmoles of HCO_3^- filtered per 24 hours in an average normal adult who has a glomerular filtration rate of 180 liters per 24 hours. Normally the kidney reabsorbs only 26 to 28 mmoles of HCO_3^- from each liter of filtrate, the so-called renal threshold for HCO_3^- under normal circumstances, and thus the kidney would be expected to rapidly excrete HCO_3^- in an uncomplicated acute metabolic alkalosis.

In chronic metabolic alkalosis the renal handling of the HCO_3^-, K^+, and Na^+ are subject to many other influences due to the fact that this condition is usually accompanied by an increased renal Na^+ reabsorption, a Cl^- depletion, and possibly K^+ depletion. The most important of these influences are the availability of the reabsorbable anion Cl^- in the glomerular filtrate, the avidity of the renal tubular Na^+ reabsorption system, and the availability of K^+.

When faced with a strong stimulus for Na^+ reabsorption, such as extracellular fluid volume depletion, the kidney responds by avid tubular reabsorption of filtered Na^+. In order to maintain electroneutrality, Na^+ reabsorption must be accompanied by the reabsorption of a reabsorbable anion or excretion of an equivalent amount of cation. The two reabsorbable anions found in any significant quantity in the glomerular filtrate are Cl^- and HCO_3^-, the concentration of Cl^- usually being about 4 times that of HCO_3^-. If there is Cl^- depletion, HCO_3^- will be vigorously reabsorbed as the accompanying anion for Na^+. Since HCO_3^- is in limited supply in the glomerular filtrate further Na^+ reabsorption will require Na^+ to be exchanged for another cation, specifically H^+ or K^+. Therefore the urine will be rich in K^+ and H^+. If K^+ depletion supervenes, even more H^+ will have to be excreted. The result of this complicated interaction of multiple factors is the production of a urine which is HCO_3^--free and acidic; thus the

(A)   (B)

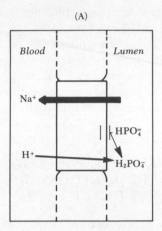 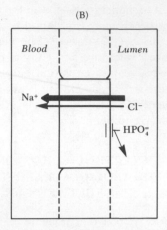

Figure 9-2. The renal response to the depletion of reabsorbable anion. *A*, Accelerated sodium reabsorption in the absence of reabsorbable anion. *B*, Accelerated sodium reabsorption in the presence of a readily reabsorbable anion (chloride). Changes in potassium secretion parallel those of hydrogen ion. (From Schwartz et al: New Engl. J. Med., *279*: 630, 1968.)

kidney not only fails to excrete the alkali excess but reabsorbs it and actually generates more HCO_3^- as it acidifies the urine in patients with a chronic metabolic alkalosis accompanied by a renal tubular avidity for Na^+ reabsorption, depletion of Cl^-, and possibly depletion of K^+. The renal response to depletion of a reabsorbable anion is depicted in Figure 9-2.

In order to correct this chronic metabolic alkalosis, Cl^- must therefore be supplied and it is also important that the volume of extracellular fluid be expanded to decrease the need for Na^+ reabsorption. NaCl infusion will fill these needs well and is used except in situations where excessive Na^+ administration would be hazardous when arginine HCl or lysine HCl are used. It has been recently demonstrated experimentally in dogs that restoration of extracellular fluid volume without increasing the filtered Cl^- load can result in selective retention of Cl^- without retention of Na^+. Thus correction of the extracellular volume deficit alone tends to make the renal response more appropriate. However, for clinical purposes it should be remembered that provision of filtered Cl^- and correction of volume depletion are the cornerstones of treatment for metabolic alkalosis along with correction of the hypokalemia (i.e., low serum potassium level) which usually exists.

As can be seen from the preceding discussion acute metabolic alkalosis will be rapidly corrected by the kidney if there is no deficit of extracellular volume, Cl^- or K^+, and the factor inciting the metabolic alkalosis is withdrawn. If any of the forementioned modifying factors are present then the alkalosis will progress and become a chronic metabolic alkalosis. Thus the renal handling of metabolic alkalosis depends more upon concomitant levels of body Cl^-, extracellular fluid volume, and K^+ than it does upon the magnitude of the noncarbonic acid deficit or alkali excess. Moreover, as previously described, there is also a variability of the ventilatory response to metabolic alkalosis depending upon the inciting agent.

In light of these multiple factors it is to be expected that the relationship between pCO_2, pH, or H^+ concentration, and $[HCO_3^-]$ of the blood is not as well defined as it is in metabolic acidosis; indeed, this is the case. There is a wide range of H^+ concentration for any given pCO_2 in metabolic alkalosis, and the usefulness of a confidence band in this type of disturbance is severely limited. Although there is some controversy over the maximal elevation of arterial CO_2 tension that can be seen as a ventilatory response to uncomplicated metabolic alkalosis, physicians working with clinical acid-base problems usually agree that an arterial pCO_2 greater than 55 mm Hg will not occur as a result of a metabolic alkalosis without some complicating factor. If values greater than this are seen, a concomitant respiratory acidosis must be considered to be present, and if the pCO_2 is greater than 50 mm Hg one should seek to rule in or out a concomitant respiratory acidosis.

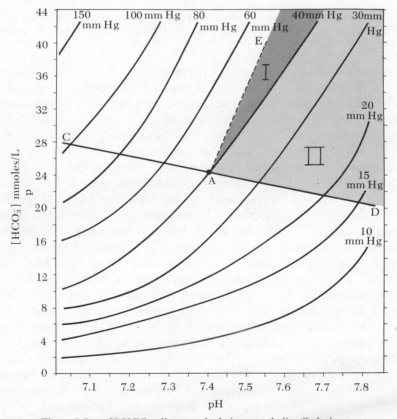

Figure 9-3. pH-HCO_3^- diagram depicting metabolic alkalosis.

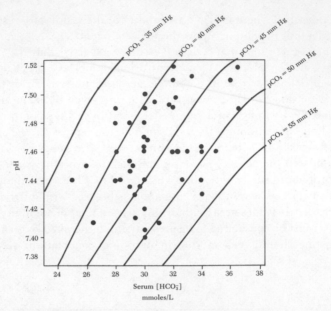

Figure 9-4. Relationship between the concentration of bicarbonate in arterial serum, the arterial blood pCO_2, and the arterial blood pH during induced metabolic alkalosis. (Modified from Goldring et al.: J. Clin. Invest., *47*:188, 1968.)

Metabolic alkalosis can also be analyzed by the pH-HCO_3^- diagram (see Fig. 9-3) but as with the case of metabolic acidosis the terminology usually used in such analysis is in apparent conflict with that primarily used in this text. Therefore it is profitable to first discuss metabolic alkalosis in the classic terms used in conjunction with the pH-HCO_3^- diagram and then to relate this terminology to that used in this text.

If a point is plotted on the diagram which represents the pH, [HCO_3^-], and pCO_2 of the true plasma of a patient and if that point is well removed from point *A* and falls within area I of Figure 9-3, the patient is probably suffering from metabolic alkalosis. If the point from a metabolic alkalosis patient falls close to the hatched line *A–E* (a line representing the maximal respiratory compensation capacity of a normal person), then the patient suffers from a metabolic alkalosis accompanied by good respiratory compensation. If on the other hand the point falls in area I near the *A–B* segment of the pCO_2 40 mm Hg isobar then the patient is said to be suffering from metabolic alkalosis with little or no respiratory compensation. Obviously from this analysis whenever the pCO_2, [HCO_3^-], and pH of true plasma lead to a point plotted in area I which is sufficiently removed from point *A*, the patient is suffering from a simple metabolic alkalosis and the closer the point falls to the *A–E* line the greater the respiratory compensation achieved. However, if the point plots in area II, i.e., somewhere between the pCO_2 40 mm Hg isobar segment *A–B* and the segment *A–D* of the nonbicarbonate buffer line, the patient is said to be probably suffering from a mixed disturbance of metabolic alkalosis plus respiratory alkalosis. The fact that the pCO_2 is below normal and the pH is elevated signals a respiratory alkalosis (see Chapter 11), and the fact that the $[HCO_3^-]$ is greater than predicted by the nonbicarbonate buffer line for a given pH and pCO_2 and the $[H^+]$ is less than normal indicates the presence of a metabolic alkalosis component.

The discussion in most of this text has differed from this classic analysis of metabolic alkalosis in only one regard; namely, the term simple metabolic alkalosis has been restricted to that area of the graph within the confidence band of which curve $A-E$ is the upper limit. Thus by this newer terminology all patients with true plasma $[HCO_3^-]$, pCO_2, and pH which plots outside of the confidence band of line $A-E$ but plots somewhere within area I or area II are designated as mixed disturbances of metabolic alkalosis plus respiratory alkalosis. However, the confidence band for line segment $A-E$ is so wide as to severely limit its usefulness and therefore no real differences exist between the two terminologies in this case. As an example of this variability of response to metabolic alkalosis refer to Figure 9-4 where pCO_2 ranges from 39 to 50 mm Hg with a $[HCO_3^-]$ of 30 mmoles/L.

REFERENCES

For Causes:

Black, D. A., and Milne, M. D.: Experimental potassium depletion in man. Clin. Sci. *11*:397 (1952).
Kassirer, J. P., Appleton, F. M., Chazan, J. A., and Schwartz, W. B.: Aldosterone in metabolic alkalosis. J. clin. Invest. *46*:10, 1558–1571 (1967).
Mulhausen, R. O., and Blumentals, A. S.: Metabolic alkalosis. Arch. intern. Med. *116*:729–739 (1965).
Roth, D. G., and Gamble, J. L.: Deoxycortisone-induced alkalosis in dogs. Amer. J. Physiol. *208*:90–93 (1965).
Womensky, R. A., and Danagh, J. H.: Potassium and sodium restriction in the normal human. J. clin. Invest. *34*:456 (1955).

For Renal Responses:

Pitts, R. F.: Physiology of the Kidney and Body Fluids. Year Book Medical Publishers Inc., Chicago, 1968.
Schwartz, W. B., and Relman, A. S.: Metabolic and renal studies in chronic potassium depletion resulting from overuse of laxatives. J. clin. Invest. *32*:258 (1953).
Schwartz, W. B., Van Ypersele de Strihou, C., and Kassirer, J. P.: Role of anions in metabolic alkalosis and potassium deficiency. New Engl. J. Med. *279*:630–639 (1968).

For Respiratory Response:

Cohen, J. J.: Selective chloride retention in repair of metabolic alkalosis without increasing filtered load. Amer. J. Physiol. *218*:165–170 (1970).
Goldring, R. M., Cannon, P. J., Heinemann, H. O., and Fishman, A. P.: Respiratory adjustment to chronic metabolic alkalosis in man. J. clin. Invest. *47*:188–202 (1968).

10

RESPIRATORY ACIDOSIS

DEFINITIONS

Respiratory acidosis is a disturbance which is primarily caused by an inability of alveolar ventilation to excrete CO_2 as rapidly as required to satisfy the existing needs of the body. The resulting CO_2 retention causes a "carbonic acid excess" relative to the acid-base needs of the body. The existing need for CO_2 excretion depends upon the rate of CO_2 production and the acid-base status of the organism. For example, let us consider the case of a person suffering from an acute metabolic acidosis of such severity that the ventilatory response required is one which reduces the arterial pCO_2 to 30 mm Hg or less. If this person maintains his arterial pCO_2 at 40 mm Hg (which is a normal value in the resting healthy male), a respiratory acidosis is considered to be present (see Chapter 8), i.e., his arterial pCO_2 is not low enough to meet the needs of his metabolic acidosis.

CAUSES OF RESPIRATORY ACIDOSIS

The common denominator in respiratory acidosis is some impairment of alveolar ventilation. Although the specific diseases which cause alveolar hypoventilation could be enumerated, a more useful approach is to outline the mechanisms leading to alveolar hypoventilation and then to cite illustrative clinical examples.

Frequency and depth of breathing, which in turn determine alveolar ventilation, are regulated by the respiratory centers in the medulla and pons.

Hypoventilation can be produced if these centers are depressed by anesthetizing agents, narcotics, sedatives, hypoxia, excessively high CO_2 concentrations, and central nervous system disease of any kind, e.g., ischemic disease, trauma, and increased cerebrospinal fluid pressure. Even with a normally functioning respiratory center, if the transmission of the neural impulse to the respiratory muscles is interrupted, alveolar hypoventilation results. This can be seen with spinal cord lesions, poliomyelitis, amyotrophic lateral sclerosis, peripheral neuritis, and myasthenia gravis or with drugs or toxins such as succinylcholine, curare, nerve gases, or botulinus toxin.

Disease of the respiratory muscles can cause alveolar hypoventilation as can diseases which severely restrict the movement of the thorax such as arthritis, scleroderma, thoracic deformities, or severe obesity. The respiratory muscles can be affected by myopathies or by generalized muscle wasting or severe hypokalemia causing weakness and thus decreased thoracic excursions. The movement of the lungs can also be restricted by intrathoracic collections such as pleural fluid, blood, or pneumothorax. In tetanus the thoracic muscle spasm secondary to increased anterior horn cell activity can result in severe restriction of chest wall motion and therefore alveolar hypoventilation.

Finally, pulmonary diseases cause alveolar hypoventilation. These diseases may be broadly divided into restrictive disorders where lung distensibility is limited, such as sarcoidosis, chronic congestive heart failure, infiltrative tumors, and atelectasis. Another category of pulmonary disorders is obstructive disease such as the chronic bronchitis-emphysema group of disorders or asthma which represent lower airway obstructive disease. Upper airway obstruction is occasionally caused by tracheal stenosis, a foreign body, or a tumor.

The most common causes of respiratory acidosis today are probably the chronic obstructive pulmonary diseases and the overuse of respiratory depressant drugs, either perscribed by a physician or taken in a suicide attempt, or more recently from accidental overdoses of self-administered narcotics such as heroin.

Frequently, multiple etiologic factors are involved, as for example in the patient with both chronic obstructive pulmonary disease and congestive heart failure who receives a narcotic as therapy for acute pulmonary edema. Another common combination is the elderly patient with chronic pulmonary disease who receives a narcotic for pain following trauma or surgery.

EXTRACELLULAR AND INTRACELLULAR BUFFERING MECHANISMS

Noncarbonic acids which in body fluids are nearly fully dissociated to H^+ and anion are slow to diffuse across cell membranes. Carbon dioxide, on the other hand, readily diffuses across cell membranes. The initial buffering of CO_2 or carbonic acid in the extracellular space yields HCO_3^-

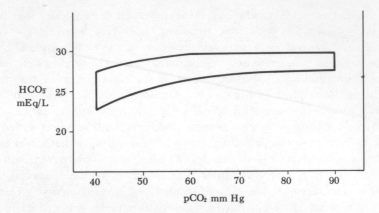

Figure 10-1. The relationship between arterial pCO_2 and arterial plasma $[HCO_3^-]$ in acute respiratory acidosis. The area enclosed within the solid lines indicates the range of $[HCO_3^-]$ seen at each pCO_2 level. (From Brackett et al.: New Engl. J. Med., *272*:12, 1965.)

in proportion to the pCO_2 increment and the available extracellular non-bicarbonate buffers (see Fig. 4-7). The concentration of HCO_3^- in the extracellular fluid increases rapidly as pCO_2 rises up to certain limits. In normal volunteers an increase in arterial pCO_2 from 40 mm Hg to 70 mm Hg gives an increase in plasma HCO_3^- from about 24 mmoles per liter to a little over 27 mmoles per liter. This is illustrated in Figure 10-1 which relates pCO_2 to HCO_3^- in acute respiratory acidosis. This is less than the expected HCO_3^- increase when whole blood is exposed *in vitro* to a CO_2 tension of 70 mm Hg.

The difference between the carbon dioxide titration curves *in vitro* and *in vivo* is due to the superior buffering ability of blood compared to interstitial fluid (compare Fig. 4-6 with Fig. 4-7). The ability of blood to buffer carbonic acid resides primarily in the hemoglobin molecule and plasma proteins. This has been discussed in detail in Chapters 3, 4, and 5.

Much of the H^+ generated when CO_2 is accumulated in extracellular fluid ultimately enters the cells and is buffered in the intracellular space. There is a consistent rise in extracellular $[K^+]$ in respiratory acidosis which occurs within several hours. The reasons for this are not completely understood. Over a period of 10 days or so, these ion shifts between extracellular fluid and intracellular fluid, especially in muscle and bone, will buffer about 25 per cent of the excess H^+ generated extracellularly by the CO_2 retention. With prolonged respiratory acidosis there is also a decreased rate of intracellular lactic acid production.

RENAL COMPENSATION IN RESPIRATORY ACIDOSIS

When arterial pCO_2 is elevated, as it is in most instances of respiratory acidosis, there is a significant increase in the ability of the kidney to reabsorb HCO_3^- which relates directly to the magnitude of the pCO_2 increase (see Fig. 10-2). This increase in HCO_3^- reabsorption occurs when the arterial pH falls and also when it is held constant in appropriately designed experiments. Thus the stimulus for the increased renal HCO_3^- conservation appears to be an increased arterial pCO_2 and not an increase in arterial H^+ concentration.

Presumably an increase in arterial pCO_2 causes an equivalent increase in the pCO_2 of the renal tubular cells thus causing an increase in the formation of carbonic acid (H_2CO_3) and an elevation of renal intracellular H^+ concentration and H^+ secretion. In addition to increasing HCO_3^- reabsorption, the increased amount of renal H^+ secretion results in *de novo* generation of HCO_3^- by the renal tubules. Simultaneously with this enhanced HCO_3^- reabsorption and generation, the kidneys increase the rate of Cl^- excretion, leading to the eventual development of hypochloremia.

This renal response to respiratory acidosis takes days to develop fully and to become the predominant factor regulating extracellular fluid $[H^+]$. Ultimately it is a great deal more effective in protecting blood pH levels than the rapidly acting buffer systems. The renal response is usually well developed by 48 hours and is most commonly complete within five days. In acute respiratory acidosis before renal compensation occurs to any significant extent, the plasma $[HCO_3^-]$ reaches about 27 mmoles per liter when the arterial pCO_2 increases to 70 mm Hg. When the renal response is complete the plasma $[HCO_3^-]$ for the same subject (i.e., arterial pCO_2 70 mm Hg) is about 37 mmoles per liter. These acute and chronic states of respiratory acidosis can be quantitatively compared by using the Henderson equation:

$$H^+ \text{ (nmoles/L)} = \frac{23.8 \text{ (mm Hg)} \times pCO_2}{HCO_3^- \text{ (mmoles/L)}}$$

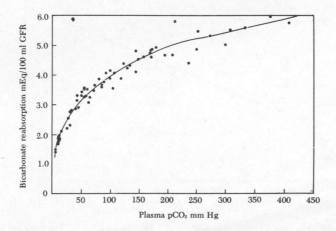

Figure 10-2. The relationship between renal HCO_3^- reabsorption and arterial plasma pCO_2. (From Rector, F. C. et al.: J. Clin. Invest., 39:1706, 1960.)

Substituting the values seen in acute respiratory acidosis before renal compensation has occurred we get:

$$H^+ = \frac{23.8 \times 70}{27} = 61.7 \text{ nmoles/L}$$

which is equivalent to a pH of 7.22. Substituting the values seen in chronic respiratory acidosis we get:

$$H^+ = \frac{23.8 \times 70}{37} = 45 \text{ nmoles/L}$$

which is equivalent to a pH of 7.35.

These calculations merely serve to illustrate the relative effectiveness of the rapid buffering mechanism which is the dominant one in the acute state and the slower but more powerful renal mechanism which is the dominant one in the chronic state. This point is further illustrated by the curves relating the pH or $[H^+]$ to pCO_2 for acute and chronic respiratory acidosis shown in Figure 10-3. The curve for the acute state relates primarily to the extracellular buffer and some of the intracellular buffer response while the curve for the chronic state relates primarily to the renal response and to a lesser extent the extracellular and intracellular buffering.

The renal response to respiratory acidosis can also have an adverse effect on patients with chronic respiratory acidosis in whom this response is complete. If these patients are treated or if there is improvement in the course of their disease resulting in an augmented alveolar ventilation, the

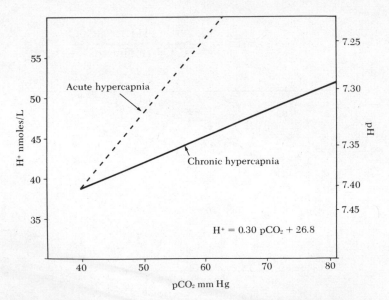

Figure 10-3. The relationship between arterial blood pCO_2 and $[H^+]$ in acute and chronic respiratory acidosis. (From Van Ypersele de Strihou, et al.: New Engl. J. Med., *275*:117, 1966.)

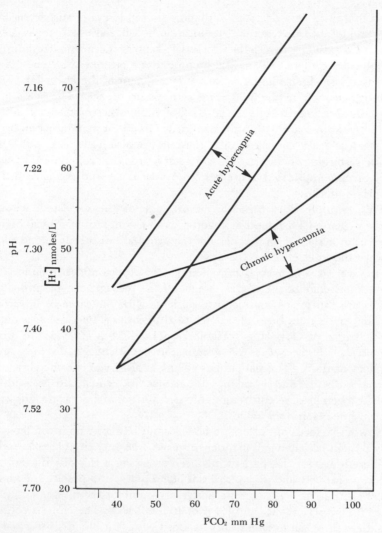

Figure 10-4. Confidence bands for acute and chronic respiratory acidosis. (Modified from Weiss and Dulfano: Ann. Int. Med., *69*:263, 1968.)

arterial pCO_2 will be rapidly reduced. They will then have an excess of HCO_3^- relative to the arterial pCO_2 level and thus will develop a metabolic alkalosis which may last several days or longer, especially if there is an extracellular Cl^- deficit such as frequently accompanies chronic respiratory acidosis (see Chapter 9).

Figure 10-4 depicts the confidence bands for uncomplicated acute and chronic respiratory acidosis using pCO_2 and $[H^+]$ or pH as the abscissa and ordinate. To illustrate the use of these confidence bands let us consider an elderly patient who is known to be in good health and not taking any medication. If this patient has upper abdominal surgery to remove his gall bladder there will be a decrease in diaphragm motion after surgery because of pain and local tissue reaction. In addition he will most likely be receiving narcotics for postoperative pain which will depress the normal responses of the respiratory centers. Let us also suppose that a postoperative chest X-ray reveals findings which suggest collapse of a segment of the lower lobe of the right lung accompanied by fluid in the right pleural space. The pCO_2 of the arterial blood is 55 mm Hg and the $[H^+]$ 50 nmoles/L equivalent to a pH of 7.30. These values are within the confidence band for acute uncomplicated respiratory acidosis, and the clinical findings are also quite compatible with an acute respiratory acidosis. Thus we can reasonably conclude that this is in fact an uncomplicated acute respiratory acidosis and we can institute appropriate therapy.

This example illustrates how knowledge of the expected acid-base response to acute CO_2 retention, together with a compatible clinical picture, enables us to make a firm diagnosis of the acid-base disturbance and directs us toward the most effective therapy.

However, it is important to remember that the confidence band alone, without knowledge of the clinical course of the patient, can be misleading. To illustrate this let us consider two patients with the same blood arterial pCO_2 and pH values but different clinical histories. The first is a patient with an arterial pCO_2 of 60 mm Hg and a pH of 7.34 or H^+ concentration of 46 nmoles/L. These values lie within the confidence band for chronic respiratory acidosis. This patient has a clinical history of chronic obstructive pulmonary disease and is taking no medication which affects acid-base balance. Therefore we can reasonably conclude that he has an uncomplicated chronic respiratory acidosis.

Now let us focus our attention on a second patient who is similar to the elderly patient who had gall bladder surgery. The only circumstances which are different are (1) this patient has been receiving a thiazide diuretic pre- and postoperatively and (2) his arterial blood pCO_2 and pH are the same as those of the patient with chronic obstructive pulmonary disease i.e., pCO_2 of 60 mm Hg, pH of 7.34, and $[H^+]$ of 46 nmoles/L. The clinical course of this patient is of acute respiratory acidosis, but the blood pCO_2 and pH values lie within the confidence band for chronic respiratory acidosis. This disparity between clinical course and blood pCO_2 and pH suggests that there may be a mixed disturbance. If we look at his medications we note that he is receiving thiazide diuretics which are a common cause of metabolic alkalosis (see Chapter 9). Thus at the time the acute respiratory acidosis developed this patient had an increase in his extracellular fluid $[HCO_3^-]$ and a decrease in his $[H^+]$ compared to the normal values. When acute CO_2 retention is superimposed upon this pre-existing acid-base status, the result is a $[H^+]$ and $[HCO_3^-]$ more like that seen in chronic respiratory acidosis than in acute respiratory acidosis.

The difference between the first and the second patient lies in the following point. The first patient's renal HCO_3^- conservation is a compensating response to primary CO_2 retention, while much of the second patient's extracellular HCO_3^- elevation is due to the renal effects of a drug. The importance of making this distinction between a patient with simple chronic respiratory acidosis and one with metabolic alkalosis and a superimposed acute respiratory acidosis lies in the therapeutic implications.

The point to be retained from the preceding discussion is that pH or $[H^+]$ and pCO_2 values which fall within a confidence band for acute or chronic respiratory acidosis do not mean that the problem must be only a simple uncomplicated disturbance. One must apply all available knowledge including the clinical history and course, any therapy given, and prior or concurrent laboratory data.

The confidence band is also useful when considering those patients whose blood pH, pCO_2, and $[HCO_3^-]$ values fall outside its limits. Let us consider a patient with advanced chronic obstructive pulmonary disease known to be suffering from chronic respiratory failure. On examination of the arterial blood, the pCO_2 is found to be 70 mm Hg, the $[H^+]$ —40 nmoles/ L, and the $[HCO_3^-]$ 42 mmoles/L. These data clearly fall outside the confidence band in Figure 10-4 for chronic respiratory acidosis and therefore we immediately suspect that he has a complicating acid-base disturbance. Since this disturbance has resulted in a $[H^+]$ lower than anticipated and a $[HCO_3^-]$ higher than expected for this degree of pCO_2 increase it must be an alkalosis. Since the $[HCO_3^-]$ is markedly elevated, the complicating disturbance cannot be a respiratory alkalosis and must therefore be a metabolic alkalosis. Obviously, the confidence bands are useful when intelligently applied and when used in conjunction with all other data available, both clinical and laboratory.

Respiratory acidosis can also be analyzed by the pH-HCO_3^- diagram. If pH, $[HCO_3^-]$, and pCO_2 can be represented by a point as a distance from point A but falling in area I of Figure 10-5, the patient probably suffers from a primary respiratory acidosis. If that point falls on the A–C segment of the nonbicarbonate buffer curve, then the patient is probably suffering from an acute respiratory acidosis with little of the H^+ generated extracellularly having entered the cells and little or no renal compensation, i.e., the extracellular buffers can fully explain the change in pH and $[HCO_3^-]$ relative to the pCO_2 change. This line segment A–C is expressed in slightly different form as the confidence band in Figure 10-1. If the point falls somewhere in area I between curve A–C and curve A–E (the curve representing maximal renal compensation in most people) then the respiratory acidosis is at least of many hours duration and involves both intracellular buffering of H^+ formed extracellularly and renal compensation for the respiratory acidosis. If the point plots somewhere on curve A–E, the renal compensation is complete and the acidosis is of reasonably long standing. To know if the patient is in a steady-state or is recovering or getting worse from this acidosis, it is necessary to plot a time course of these points as explained in Chapter 8 for metabolic acidosis. Mixed problems involving

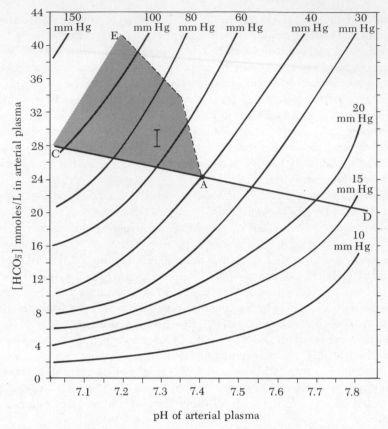

Figure 10-5. Analysis of respiratory acidosis by the pH-HCO$_3^-$ diagram.

metabolic acidosis and respiratory acidosis relative to the pH-HCO$_3^-$
diagram were discussed in Chapter 8.

There is an interesting change in the slope of the line segment *A–E*
at pH 7.31 and [HCO$_3^-$] of 35 mmoles/L. This change in slope reflects the
fact that the renal bicarbonate conserving mechanism seems to resist further
increases from a pCO$_2$ stimulus at a pCO$_2$ of about 70 mm Hg and [HCO$_3^-$]
of 35 to 37 mmoles/L. From that point on the rise in extracellular fluid
[HCO$_3^-$] per mm Hg pCO$_2$ increase is little different than the slope of line
segment *A–C* which represents HCO$_3^-$ generated by nonbicarbonate buffering of H$^+$. The clinical correlate to this, according to many physicians
dealing with acid-base problems, is that a [HCO$_3^-$] above 40 mmoles/L
in a patient with chronic respiratory acidosis usually represents either HCO$_3^-$
administration or a complicating metabolic alkalosis. The pCO$_2$-H$^+$ (or pH)
confidence bands and the pH-HCO$_3^-$ diagram are merely different ways of
looking at the same data. This is nicely illustrated by comparing the change
in slope of *A–E* in Figure 10-5 with the abrupt change in the slope of the
confidence band for chronic respiratory acidosis in Figure 10-4 which also

occurs at a pCO_2 of about 70 mm Hg and reflects the abrupt decrease in renal HCO_3^- conservation capacity.

Although the confidence band expressed as pCO_2 and $[H^+]$ may be more useful when dealing with patients who have acid-base disorders, the pH-HCO_3^- diagram provides information about the different components of the total response to a CO_2 load that is not as easily derived from other means of graphic display.

REFERENCES

For Causes of Respiratory Acidosis:

Comroe, J. H., Forster, R. E., Dubois, A. B., Briscoe, W. A., and Carlsen, E.: The Lung: Clinical Physiology and Pulmonary Function Tests. Year Book Medical Publishers, Inc., Chicago, 1962.

For Extracellular Buffering Mechanisms:

Tenney, S. M., and Lamb, T. W.: Physiological Consequences of Hypoventilation and Hyperventilation. Chapter 37, Handbook of Physiology, Sect. 3, Respiration, Vol. II., Amer. Physiol. Soc. Washington, D.C., 1965.

For Chronic and Respiratory Acidosis in Man:

Brackett, N. C., Cohen, J. J., and Schwartz, W. B.: Carbon dioxide titration curve of normal man. New Engl. J. Med., 1965, Vol. 272, 6-12.
Van Ypersele de Strihou, C., Brasseur, L., and DeConinck, J.: The "carbon-dioxide response curve" for chronic hypercapnia in man. New Engl. J. Med. *275*:117–122 (1966).
Weiss, E. B., and Dulfano, M. J.: Quantitative acid-base dynamics in chronic pulmonary disease: defense of pH during acute respiratory acidosis superimposed upon chronic hypocapnia. Ann. intern. Med. *69*:263–272 (1968).

For Renal Compensation of Respiratory Acidosis:

Pitts, R. S.: Physiology of the Kidney and Body Fluids, Year Book Medical Publishers, Inc., Chicago, 1968.
Refsum, H. E.: Acid-base status in patients with chronic hypercapnia and hypoxemia. Clin. Sci. *2*:407–415 (1964).

11

RESPIRATORY ALKALOSIS

DEFINITIONS

Respiratory alkalosis is defined as alveolar ventilation in excess of the existing need of the body for CO_2 elimination resulting in an arterial pCO_2 lower than appropriate for the circumstances. Such excess in alveolar ventilation is called hyperventilation. Increase in alveolar ventilation which precisely meets a physiologic need is not considered hyperventilation. Hyperpnea is the general term which describes any increase in ventilatory effort.

CAUSES

Alveolar hyperventilation may be caused by a variety of factors. An interesting phenomenon is the "hyperventilation syndrome" in which chronic or recurrent alveolar hyperventilation occurs for no apparent reason. The patients are anxious and tense and often complain of a number of symptoms, especially numbness and tingling of the extremities and circumoral area, sensations of light-headedness, and flexion spasms of the wrists. They also develop electroencephalographic (EEG) abnormalities. Some of the symptoms appear to develop concurrently with the EEG changes and may be related to the extent and rate of fall in arterial pCO_2. Other symptoms, such as tetany, do not correlate well with EEG changes but may correlate with the duration of the low arterial pCO_2 levels. Healthy individuals, when subjected to marked reductions in arterial pCO_2 induced by voluntary overbreathing, show considerably less symptomatology than patients suffering from "hyperventilation syndrome."

Alveolar hyperventilation may also be produced by drugs. This is frequently observed clinically with salicylate overdosage which simultaneously induces a metabolic acidosis. Excessive doses of analeptic drugs, 2-4 dinitrophenol, and paraldehyde have also been known to produce hyperventilation. Hormones such as epinephrine and progesterone increase ventilation, presumably by acting upon the central nervous system. Lesions of the central nervous system including meningitis, encephalitis, cerebral hemorrhage, or trauma are known to increase alveolar ventilation. Fever, shock, gram-negative bacteremia without fever or shock, and hyperthyroidism, all may cause hyperventilation. Almost any interstitial pulmonary disease can lead to hyperventilation, e.g., sarcoidosis and pulmonary fibrosis. Hepatic failure is often accompanied by a respiratory alkalosis.

When a patient's ventilatory efforts are being assisted by a mechanical ventilator, with the patient setting his own rate of breathing, alveolar hyperventilation occasionally occurs. If the patient's ventilation is being completely controlled, with the ventilator setting both the rate and depth of ventilatory effort, alveolar hyperventilation is even more likely to occur.

When anoxemia is present there is an increase in alveolar ventilation, which is accomplished primarily by an increase in the tidal volume, with little increase in the frequency of breathing. This response is initiated in the carotid and aortic body chemoreceptors, which appear to sense changes in the arterial pO_2, and does not become significant until the concentration of oxygen in inspired air at sea level is reduced from 21 to 16 per cent. Although not sensitive, this chemoreceptor mechanism is persistent, as is well demonstrated by people living at high altitudes.

Increased alveolar ventilation is a normal response to metabolic acidosis (see Chapter 8). In patients who have had metabolic acidosis for more than 24 hours, the increased alveolar ventilation may persist for several hours to a day or more after correction of the metabolic acidosis. Since the continued increase in alveolar ventilation is no longer appropriate for the new acid-base state of the patient, this hyperventilation may cause a respiratory alkalosis, which is usually mild.

EXTRACELLULAR AND INTRACELLULAR BUFFERING

During the first three minutes of acute alveolar hyperventilation most of the CO_2 excreted comes from the pool of CO_2 that was already in the alveolar air prior to the hyperventilation. After this time, the expired CO_2 is almost entirely derived from pools of CO_2 in the blood and tissue water so that the total body CO_2 content is rapidly depleted.

The arterial blood pH begins to rise in 15 to 20 seconds after hyperventilation begins and becomes maximal in 10 to 15 minutes. The plasma HCO_3^- level drops along a similiar time course. A drop of pCO_2 from 45 mm Hg to 15 mm Hg will result in a fall in plasma $[H^+]$ to about 18 to 25 nmoles (a pH of 7.61 to 7.74) and a $[HCO_3^-]$ of 15 to 20 mmoles/L.

About 20 per cent of the extracellular H^+ deficit caused by acute respiratory alkalosis is in time replenished by H^+ from the intracellular fluid. Simultaneously there is an entry of Na^+ and K^+ into the intracellular fluid from the extracellular space. There is also an increase in blood lactic acid. The increased production of lactic acid appears to be due to a combination of factors. These include the decreased efficiency in the delivery of O_2 to the tissues (because of the decreased ability of hemoglobin to release O_2 as the pH increases), peripheral vasoconstriction, epinephrine release, and a direct effect of the partial pressure of carbon dioxide upon lactic acid metabolism.

When the arterial pCO_2 in healthy young people is decreased and maintained at levels of 15 to 20 mm Hg for 3 hours, the extracellular fluid alkalosis appears to be accompanied by a proportional intracellular fluid alkalosis with loss of HCO_3^- from intracellular as well as extracellular water.

RENAL RESPONSE TO RESPIRATORY ALKALOSIS

While buffering and intracellular-extracellular ion shifts occur over a period ranging from minutes to hours, the renal response is not a major factor early in respiratory alkalosis but becomes so if this state persists for several days or longer. In acute respiratory alkalosis, there is a decrease in urine H^+ excretion and an increase in HCO_3^- excretion in the urine. The stimulus for the decreased HCO_3^- reabsorption appears to be related to the decreased extracellular fluid pCO_2. There is also an increased excretion of K^+ in the urine. It should be noted that these responses to respiratory alkalosis may be altered in the patient in whom Na^+ and Cl^- is depleted.

VARIOUS COURSES OF ACUTE AND CHRONIC
RESPIRATORY ALKALOSIS

The studies on people with chronic respiratory alkalosis, i.e., alveolar hyperventilation lasting longer than five days, have all been performed at high altitude and therefore may have been influenced by the fact that these people are chronically hypoxemic. This chronic hypoxemia causes the alveolar hyperventilation, but it also causes a considerable increase in the hemoglobin content of the blood and a shift in the ratio of oxygenated to nonoxygenated hemoglobin. These subjects have had time to effect a complete renal response to their respiratory alkalosis. It appears that for the same level of pCO_2, there is a correspondingly lower arterial pH (i.e., higher arterial $[H^+]$) in chronic respiratory alkalosis than in steady-state acute respiratory alkalosis. For example, in a person living at a high altitude

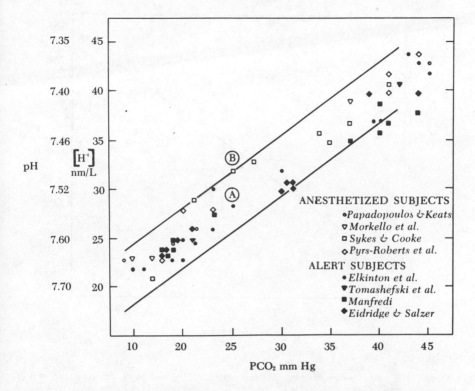

Figure 11-1. The confidence band for the acid-base response of man to acute hypocapnia. The symbols refer to subjects studied by the authors noted in the figure. (Modified from Arbus et al.: New Engl. J. Med., *280*:117, 1969.)

for five days and having a pCO_2 of 25 mm Hg the pH is usually in the neighborhood of 7.45, or the $[H^+]$ is 37 nmoles per liter. In contrast, a person with acute respiratory alkalosis whose pCO_2 is 25 mm Hg usually has a pH of about 7.56 or a $[H^+]$ of about 27 nmoles per liter.

In considering respiratory alkalosis from a clinical point of view we can use the confidence band shown in Figure 11-1 which was derived from people who were over-ventilated artificially (Arbus et al.) and who reached fairly stable pCO_2, pH, and HCO_3^- values in their arterial blood after about 10 minutes. These values remained fairly steady during the next 120 minutes of the study period. The confidence band was derived from the slope of the prediction line which is illustrated in Figure 11-2 and includes 95 per cent of the $[H^+]$ values to be expected for any given pCO_2 value in uncomplicated acute respiratory alkalosis. Figure 11-3 is the same confidence band with the variables expressed as arterial pCO_2 and $[HCO_3^-]$.

The use of these confidence bands is best illustrated by several examples. Let us consider a patient with an arterial pCO_2 of 25 mm Hg, $[H^+]$ of 28 nmoles/L, and $[HCO_3^-]$ of 21 mmoles/L. This patient is represented by

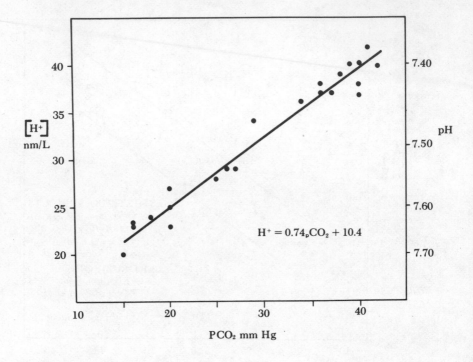

Figure 11-2. Steady-state relation between arterial $[H^+]$ and pCO_2 during acute hypocapnia in man. (From Arbus et al.: New Engl. J. Med., *280*:117, 1969.)

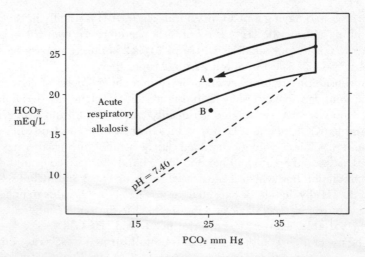

Figure 11-3. The confidence band for the steady-state relation between arterial pCO_2 and $[HCO_3^-]$ in acute hypocapnia. (From Arbus et al.: New Engl. J. Med., *280*:117, 1969.)

point A on Figures 11-1 and 11-3. If this patient has any one of the conditions listed earlier as causes of respiratory alkalosis and nothing in his clinical history to suggest a complicating acid-base disturbance, we can assume that he has an acute uncomplicated respiratory alkalosis.

Now consider another patient whose blood values lie at point B on Figures 11-1 and 11-3, i.e., pCO_2 of 25 mm Hg, $[HCO_3^-]$ of 17 mmoles/L, and $[H^+]$ of 33 nmoles/L. Point B is clearly outside the confidence band for uncomplicated respiratory alkalosis. The $[H^+]$ is higher and the $[HCO_3^-]$ lower than would be expected for a pCO_2 of 25 mm Hg. This leads us to the conclusion that the $[H^+]$ excess and the $[HCO_3^-]$ lack is caused by a co-existing acid-base disturbance which by definition is an acidosis. Since respiratory alkalosis and respiratory acidosis are both dependent upon the level of alveolar ventilation they cannot co-exist in the same patient. Therefore the complicating acid-base disturbance must be a metabolic acidosis. If the clinical history reveals that this patient ingested toxic amounts of aspirin we can conclude that he is suffering from a mixed respiratory alkalosis and metabolic acidosis.

Although values that lie outside the confidence band, such as those at point B in Figures 11-1 and 11-3, indicate a mixed acid-base disturbance, values that fall within the confidence band (e.g., point A) do not necessarily indicate the presence of an uncomplicated acute respiratory alkalosis.

To illustrate this, consider a patient with metabolic acidosis due to diabetes mellitus who has received intravenous HCO_3^- as therapy. An abnormally elevated level of alveolar ventilation may persist in such patients for hours or days. Thus the arterial pCO_2 may be held at an inappropriately low level for the partially corrected metabolic acidosis causing the $[H^+]$ of the extracellular fluid to fall to alkaline levels. This patient is indistinguishable from one with an uncomplicated respiratory alkalosis if only the blood pCO_2, $[HCO_3^-]$, and $[H^+]$ values are considered. The clinical history of diabetic acidosis and the knowledge of HCO_3^- therapy are essential to the correct interpretation and therapy of this patient's acid-base disturbance. The fact that this patient has an alkalemia (i.e., alkaline pH of blood) does not mean that an acidosis (see definitions of Chapter 8) cannot be present. In this instance the acidosis has been partially corrected by a therapeutically induced metabolic alkalosis due to HCO_3^- infusion. The persistence of alveolar hyperventilation at a level which is now inappropriate has added a component of acute respiratory alkalosis causing a frankly alkaline extracellular fluid pH. The importance of making the distinction between this patient's complex acid-base disturbance and a simple acute respiratory alkalosis resides in the recognition that the underlying diabetic acidosis is probably still present and needs further therapy.

Respiratory alkalosis can also be analyzed by the pH-HCO_3^- diagram. If the pH, $[HCO_3^-]$, and pCO_2 measurements can be represented by a point at a distance from point A but falling in area I of Figure 11-4, the patient probably suffers from a primary respiratory alkalosis, but of course a definitive diagnosis requires a consideration of the entire clinical picture. If

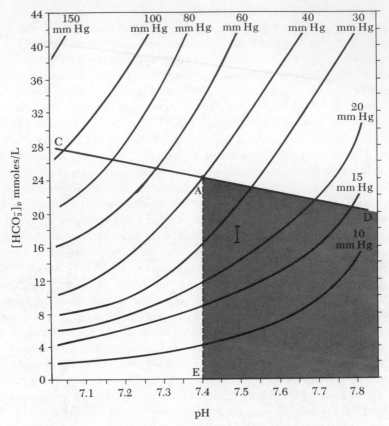

Figure 11-4. Analysis of respiratory alkalosis by the pH-HCO$_3^-$ diagram.

that point falls on the *A–D* segment of the nonbicarbonate buffer curve, then the patient is probably suffering from an acute respiratory alkalosis with the extracellular fluid being little influenced by either the intracellular buffer systems or renal compensation, i.e., the extracellular buffers can fully explain the change in pH and [HCO$_3^-$] relative to the pCO$_2$ change. If the point falls somewhere in area I between curve *A–D* and curve *A–E* then the respiratory alkalosis is of sufficient duration to involve both the intracellular buffers and renal compensation of the respiratory alkalosis. Since in many cases the pH of the blood can become normal during chronic respiratory alkalosis, i.e., curve *A–E* can be reached, it is obvious that intracellular buffering combined with renal compensation can totally readjust the [H$^+$] of the extracellular fluid to normal in the chronic steady-state. To know if the patient is in a steady-state or is recovering or getting worse from this alkalosis, it is necessary to plot a time course of these points as explained in Chapter 8 for metabolic acidosis. Mixed problems involving metabolic alkalosis and respiratory alkalosis relative to the pH-HCO$_3^-$ diagram were discussed in Chapter 9.

The course of chronic respiratory alkalosis has not been studied outside of high altitude studies as previously mentioned, so that a confidence band for uncomplicated chronic respiratory alkalosis is not available. However, as a general rule it should be anticipated that for the same degree of pCO_2 decrement, the $[H^+]$ decrement (or pH increase) will be less than that seen in acute respiratory alkalosis.

REFERENCES

For Causes:

Comroe, J. H., Forster, R. E., Dubois, A. B., Briscoe, W. A., and Carlsen, E.: The Lung: Clinical Physiology and Pulmonary Function Tests. Year Book Medical Publishers, Inc., Chicago, 1962.
Eichenholz, A.: Respiratory alkalosis. Arch. intern. Med. *116*:699 (1965).
Engel, G. L., Ferris, E. B., and Logan, M.: Hyperventilation: analysis of clinical symptomatology. Ann. intern. Med. *27*:5 (1947).
Saltzmann, H. A., Heyman, A., and Sieker, H. O.: Correlation of clinical and physiologic manifestations of sustained hyperventilation. New Engl. J. Med. *268*:1431–1436 (1963).

For Extracellular and Intracellular Buffering:

Geivisch, G., Berger, L., and Pitts, R.: The extra-renal response to acute acid-base disturbances of respiratory origin. J. clin. Invest. *34*:231–245 (1955).
Manfredi, F.: Effects of hypocapnia and hypercapnia on intracellular acid-base equilibrium in man. J. Lab. clin. Med. *69*:304–312 (1967).
Tenney, S. M., and Lamb, T. W.: Physiological Consequences of Hypoventilation and Hyperventilation. Chapter 37, Handbook of Physiology, Sect. 3, Respiration, Vol. II., Amer. Physiol. Soc., Washington, D.C., 1965.
Tomashefski, J. F., Carter, E. T., and Lipsky, J. A.: Carbon dioxide and acid-base transients during hyperventilation. J. appl. Physiol. *17*:228–232 (1962).

For Renal Response:

Barker, E. S., Singer, R. B., Elkinton, J. R., and Clark, J. K.: Renal response in man to acute experimental respiratory alkalosis and acidosis. J. clin. Invest. *36*:515–529 (1957).
Pitts, R. F.: Physiology of the Kidney and Body Fluids. Year Book Medical Publishers, Inc., Chicago, 1968.

For Various Courses of Respiratory Alkalosis:

Arbus, G. S., Hebert, L. A., Levesque, P. R., Etsten, B. E., and Schwartz, W. B.: Characterization and clinical application of the "significance band" for acute respiratory alkalosis. New Engl. J. Med. *280*:117–123 (1969).
Dill, D. B., Talbott, J. H., and Consolazio, W. V.: Blood as a physico-chemical system. Man at high altitudes. J. biol. Chem. *118*:649–666 (1937).
Dill, D. B., Talbott, J. H., and Consolazio, W. V.: Blood as a physico-chemical system. Man at high altitudes. J. Biol. Chem. *118*:649–666(1937).
Schwartz, W. B.: Characterization and clinical application of the "significance band" for acute respiratory alkalosis. New Engl. J. Med. *280*:117–123(1969).

12

THE CLINICAL APPLICATION OF THE PRINCIPLES OF ACID-BASE REGULATION

When the physician is confronted with a clinical problem involving a derangement of acid-base balance, all available information must be utilized. This includes the principles discussed in the preceding eleven chapters, the clinical history and examination of the patient, and other relevant data which will be briefly mentioned at this time.

The blood plasma contains both cations and anions. When the concentration of these ions is expressed as milliequivalents per liter of plasma, the total number of mEq of the cations must equal that of the anions. The concentration in venous plasma of the major cations and anions are: Na^+, 135–145 mEq/L; K^+, 3.5–5.0 mEq/L; Cl^-, 95–108 mEq/L; and HCO_3^-, 24–28 mEq/L.

Subtracting the sum of mEq of Cl^- and HCO_3^- per liter of plasma from the mEq of Na^+ per liter provides a rough estimate of the mEq of anions other than Cl^- and HCO_3^-. This value is referred to as the undetermined anion concentration or the "anion gap" or Δ (delta). If the "anion gap" is greater than 12–14 mEq/L, the concentration of undetermined anion in the plasma is greater than normal. An "anion gap" of less than 5 mEq/L is extremely unlikely and probably is due to laboratory error. These undetermined anions consist of sulfate, phosphates, polyanionic plasma proteins, and anions of organic acids. Most of these anions are usually the product of metabolic processes which generate H^+ (see Chapter 6). The presence of an increased "anion gap" usually indicates an excess of H^+ derived from non-carbonic acid; this obviously occurs in metabolic acidosis but may also result from a compensatory increase in the production of lactic acid in response to respiratory alkalosis (see Chapter 11). In respiratory acidosis the "anion gap" is not increased since the excess H^+ is the result of an increase in the "carbonic acid pool" rather than the fixed or noncarbonic acid pool.

Following therapy, or in instances of mixed or complicated disturbances, the pH or $[H^+]$ of the plasma may be normal despite the presence of an increased concentration of undetermined anions. Thus, calculation of the Δ can be quite useful in the analysis of the complicated problems which are not infrequently encountered clinically.

The blood urea nitrogen (BUN) and serum creatinine concentrations provide information about the adequacy of renal function. Elevation of their levels indicates a significantly diminished renal function. The normal BUN is 11–18 mg/100 ml of blood, and the creatinine normally is 0.8–1.4 mg/100 ml of serum. Since the diet of western man results in an excessive production of noncarbonic acid over alkali of about 40–60 mEq of H^+ per 24 hours, failure of renal function will eventually result in metabolic acidosis which may not be reflected by an increase in the blood $[H^+]$ if another complicating acid-base disturbance co-exists.

With this added information in mind, let us use the information developed in this book to analyze the clinical course of several patients with acid-base disturbances. One method of approaching clinical acid-base problems is to use the composite diagram shown in Figure 12-1 which is a combination of the confidence bands illustrated by figures presented in Chapters 8, 10, and 11 with the addition of an approximate confidence band for metabolic alkalosis. The hatched lines represent the HCO_3^- isopleths* expressed as mmoles/L. The use of this diagram cannot serve as a substitute for the careful consideration of the clinical features of the patient's illness, the prior and current therapy, and other parameters such as the BUN, creatinine, "anion gap," and serum electrolytes (Na^+, K^+, and Cl^-). It can, however, serve as a powerful tool in analyzing simple or complex acid-base disturbances.

Case A, presented in the diagram of Figure 12-1*A*, is that of a 24-year-old female who was admitted to the hospital with a history of frequent urination, excessive thirst, and abdominal discomfort which had been present for three days. At the time of admission her skin was warm and dry, her breathing

---

* *Isopleths*: The line connecting points on a graph that have equal or corresponding values with regard to certain variables is called an isopleth. In the case of the diagram in this chapter, $[HCO_3^-]$ in mmoles/L is the variable.

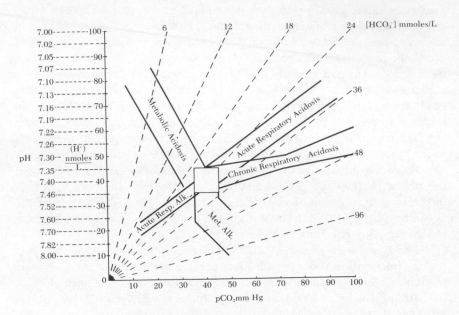

Figure 12-1. Composite acid-base diagram.

rapid and deep. She was drowsy but coherent. Her blood glucose concentration was 625 mg/100 ml blood, and her serum tested strongly positive for acetone when diluted three times with water. There was no history of any previous ailment and the patient had not been taking any medication. Point 1 on Figure 12-1A represents the values of pCO_2, $[H^+]$, and $[HCO_3^-]$ in arterial blood when the patient was admitted to the hospital: the $[H^+]$ is 110 nM, the pCO_2 is 17 mm Hg, and the $[HCO_3^-]$ is 3.5 mM.

At this point in time the patient has a severe acute metabolic acidosis but the pCO_2 is slightly higher than would be expected for this degree of H^+ increase if the confidence band is extrapolated to a H^+ concentration of 110 nM. Accordingly, a diagnosis of diabetic keto-acidosis was made and therapy was begun, namely, insulin, intravenous fluids, NaCl, and HCO_3^-.

The data obtained about 1 hour after therapy was started is plotted as point 2 on Figure 12-1A: the $[H^+]$ is 73 nM, the pCO_2 is 13 mm Hg, and the $[HCO_3^-]$ is 4.5 mM. At this point the acidosis is slightly improved and the laboratory values now fall within the confidence band for uncomplicated acute metabolic acidosis. It should be noted that the pCO_2 is lower than it was at the time the patient was admitted to the hospital and the $[H^+]$ is not as high as on admission. This may appear paradoxical but it is understandable when it is realized that the H^+ stimulus to ventilation in metabolic acidosis is greatest at a pH of about 7.10 with no increase and possibly a decrease in ventilatory effort if the pH drops below 7.10 (see Chapter 8).

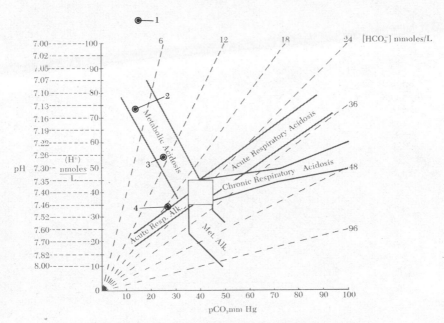

Figure 12-1A. Composite acid-base diagram for Case A.

Therefore the adjustment of pH towards normal was largely due to the improved ventilation.

The data at point 3 (Fig. 12-1A) represents the values in arterial blood six hours after therapy was begun: the [H$^+$] is 53 nM, the pCO$_2$ is 24 mm Hg, and the [HCO$_3^-$] is 11 mM. These values are well within the confidence band for acute uncomplicated metabolic acidosis and at this point the patient is responding to further therapy with insulin and fluids containing NaCl and K$^+$.

Point 4 represents the laboratory acid-base data 24 hours after admission: the [H$^+$] is 33 nM, the pCO$_2$ is 26 mm Hg, and the [HCO$_3^-$] is 19 mM. This patient now has a mild degree of respiratory alkalosis due to the continued increase in alveolar ventilation which may inappropriately remain for several hours or days while the patient is recovering from metabolic acidosis.

This case illustrates an example of a clinical and laboratory picture which is compatible with and falls within the confidence band for an acute uncomplicated disturbance of acid-base balance. Although there were several unusual features about this case the course basically represents an uncomplicated acute derangement of acid-base balance which is easily understood by following the serial determinations of [H$^+$], pCO$_2$, and [HCO$_3^-$] plotted on the acid-base diagram of Figure 12-1.

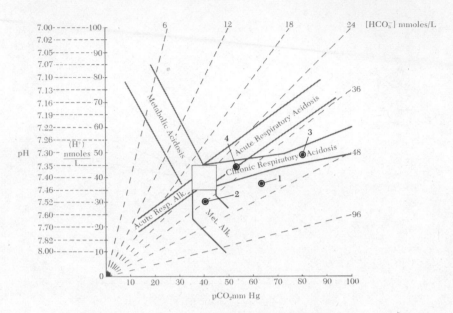

Figure 12-1B. Composite acid-base diagram for Case B.

Case B (Fig. 12-1*B*) is that of a 61-year-old man with chronic ventilatory insufficiency due to old pulmonary tuberculosis. He was readmitted to the hospital because of increasing shortness of breath and a temperature of 100°F which had been present for three days prior to admission. X-rays revealed that pneumonia, which had not been present in previous films, had developed in the lower lobe of the right lung. On the left, there was evidence of a bronchopleural fistula, thickened pleura, and collapse of the left lung, a condition which had been present for years. It had previously been determined that the left lung had no useful ventilatory function. Therapy prior to admission had included a low salt diet, thiazide diuretics, and digitalis. The patient had not received adrenal steroid hormones and had not vomited. Point 1 on Figure 12-1*B* represents the arterial blood pCO_2, $[H^+]$, and $[HCO_3^-]$ at the time of hospital admission: the $[H^+]$ is 37.5 nM, the pCO_2 is 64 mm Hg, and the $[HCO_3^-]$ is 42 mM. The other laboratory data were as follows: creatinine, 1.1 mg/100 ml; BUN, 11 mg/100 ml; CO_2 content,* 44 mmoles/L; Cl^-, 89 mEq/L; Na^+, 146 mEq/L; and K^+, 3.6 mEq/L. At this point in time the knowledge of the

---

* CO_2 content represents the venous serum concentration of H_2CO_3, CO_2, and HCO_3^- as a sum total, of which HCO_3^- usually comprises 95 per cent. (see Chapter 4 for further discussion of CO_2 content). If both pCO_2 and $[HCO_3^-]$ are known, the CO_2 content can be easily calculated. However, all of these data are presented here because clinical chemistry laboratories usually present the data as CO_2 content determined in venous serum rather than as arterial $[HCO_3^-]$ and pCO_2.

clinical history and examination of the laboratory data with reference to the diagram delineates the problem fairly clearly. This patient represents an example of chronic respiratory acidosis with some degree of metabolic alkalosis because of diuretic therapy and low salt diet.

Point 2 is the acid-base data obtained on the day of admission after four hours of assisted ventilation in a respiratory care unit: the [H$^+$] is 27 nM, the pCO$_2$ is 40 mm Hg, and the [HCO$_3^-$] is 36 mM. This patient now has a metabolic alkalosis since his pCO$_2$ has been rapidly reduced and his HCO$_3^-$ concentration has remained elevated because there was insufficient time for renal excretion of HCO$_3^-$. However, if sufficient time had elapsed for such a renal response it is likely that he would remain in a state of metabolic alkalosis because the extracellular volume and Cl$^-$ deficit induced by his low salt diet and diuretic therapy prior to admission would not permit the kidney to excrete sufficient HCO$_3^-$ (see Chapter 9).

Point 3 (Fig. 12-1B) represents the [H$^+$], [HCO$_3^-$], and pCO$_2$ six hours after the previous data set forth in point 2. During this time the settings on the ventilator were altered to allow some degree of retention of CO$_2$, and the patient was given small doses of morphine to better enable him to cooperate with the ventilator. The following laboratory data are the results: the [H$^+$] is 48 nM, the pCO$_2$ is 79 mm Hg, and the [HCO$_3^-$] is 41 mM. If we consider the laboratory values as an isolated set of data it would appear that this patient now has an uncomplicated chronic respiratory acidosis. However, we know that he has some degree of metabolic alkalosis which has not been corrected. We also know that his pCO$_2$ at this time is 15 mm Hg higher than the presumably steady state value obtained on admission some ten hours earlier and that he has received morphine sulfate which depresses the respiratory centers of the medulla. This patient who has chronically suffered from a respiratory acidosis with some degree of metabolic alkalosis now has, in addition, an acute respiratory acidosis caused by the inappropriately low alveolar ventilation selected by his therapists.

Point 4 represents the acid-base data four days after admission: the [H$^+$] is 39 nM, the pCO$_2$ is 50 mm Hg, and the [HCO$_3^-$] is 31 mM. During these four days frequent adjustments of ventilator settings were made and Cl$^-$ replacement therapy instituted. The picture is now one of a moderate degree of uncomplicated chronic respiratory acidosis. This case actually represents a clinical course frequently encountered in individuals who have chronic respiratory failure and develop an acute worsening of their ventilatory status with the result that they must be admitted to a hospital. The patient usually presents with a chronic respiratory acidosis and a metabolic alkalosis because of diuretics and a low salt diet. In this particular case, the metabolic alkalosis was evident when one looked at the admission laboratory data and compared them with the confidence band for chronic respiratory acidosis. Frequently, these patients receive assisted or controlled ventilation and are too enthusiastically ventilated with the result that the respiratory acidosis is corrected and the underlying metabolic alkalosis becomes quite

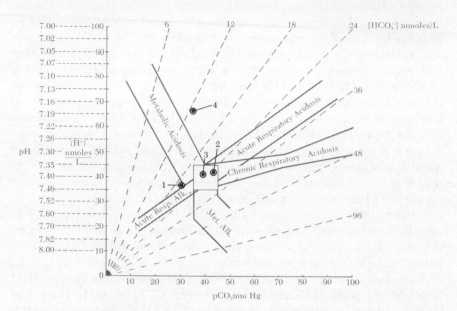

Figure 12-1C. Composite acid-base diagram for Case C.

evident. Following this, as is demonstrated by point 3, the ventilator settings are changed and the patient is frequently underventilated until finally the proper degree of ventilation is achieved and the patient settles into his usual state of chronic respiratory acidosis.

In this instance the diagram was helpful in pointing out that at the time of admission the patient's condition was not an uncomplicated chronic respiratory acidosis and that a metabolic alkalosis was present. In addition, we were also alerted to the fact that at point 3 in the patient's course, an acute respiratory acidosis was superimposed upon his chronic respiratory acidosis.

Case C (Fig. 12-1*C*) is that of a 42-year-old female admitted to the hospital because of nausea and vomiting present for one month prior to admission. At first, this vomiting was only occasional, but during the 10 days prior to admission the patient had vomited all solid food, and for 48 hours prior to entering the hospital she had vomited all liquids as well. Two years before, this patient had been diagnosed as having progressive systemic sclerosis (scleroderma). This disease can cause a rapidly progressive renal failure. Her only known therapy was two buffered aspirin tablets four times a day. Laboratory data on the day of admission are represented at point 1: the $[H^+]$ was 35 nM, the pCO_2 was 29 mm Hg, and the $[HCO_3^-]$ was 20 mM. Other data are: creatinine, 4 mg/100 ml, BUN, 52 mg/100 ml; Na^+, 137 mEq/L; K^+, 3.7 mEq/L; Cl^-, 96 mEq/L; CO_2 content, 22 mmoles/L; and Δ, 19 mEq/L.

Even though the arterial blood $[H^+]$, pCO_2, and $[HCO_3^-]$ are compatible with acute respiratory alkalosis, the clinical picture directs us towards a mixed metabolic acidosis and metabolic alkalosis. The history of severe prolonged vomiting and the lowered serum chloride strongly suggest that a metabolic alkalosis is present, and indeed the blood pH is slightly elevated. However, there is also an elevated BUN and creatinine and the "anion gap" is 19 mEq/L. In addition, the pCO_2 is decreased, which is not the case in an uncomplicated metabolic alkalosis. With this information we must conclude that she has a mixed metabolic acidosis and metabolic alkalosis due to renal failure and prolonged loss of gastric contents from the vomiting. Moreover, the data suggest but do not unequivocally establish the concomitant presence of respiratory alkalosis.

Point 2 represents the laboratory data six days after admission during which time the patient had steadily continued to vomit and had received no diuretics, adrenal steroids, or HCO_3^- therapy. At this time the $[H^+]$ was 40.5 nM, the pCO_2 was 40 mm Hg, the $[HCO_3^-]$ was 25 mM, and the CO_2 content was 28 mmoles/L, all within normal limits, but the further rise in BUN, creatinine, and "anion gap" (creatinine, 6.3 mg/100 ml; BUN, 79 mg/100 ml; Na^+, 134 mEq/L; K^+, 3.4 mEq/L; Cl^-, 83 mEq/L; and Δ, 23 mEq/L) indicate worsening of the metabolic acidosis which is matched in terms of $[H^+]$ by the metabolic alkalosis from the continued vomiting. If one had considered only the arterial blood pH, pCO_2, and $[HCO_3^-]$, the true picture of this patient's acid-base status would not have emerged.

Point 3 represents the laboratory data 12 days after admission: $[H^+]$, 39 nM; pCO_2, 38 mm Hg; $[HCO_3^-]$, 24 mM; creatinine, 12.2 mg/100 ml; BUN, 93 mg/100 ml; Na^+, 129 mEq/L; K^+, 4.4 mEq/L; Cl^-, 75 mEq/L; CO_2 content, 26 mmoles/L; and Δ, 28 mEq/L. The patient had not vomited for six days prior to this blood sample. At this point the blood chemistry resembles very much that of the previous study with the exception that the degree of renal failure has worsened as evidenced by the continuing increase in the BUN, creatinine, and "anion gap."

Point 4 (BUN, 130 mg/100 ml; creatinine, 23.6 mg/100 ml; $[H^+]$, 66 nM; pCO_2, 32.5 mm Hg; $[HCO_3^-]$, 12.7 mM; Na^+, 128 mEq/L; K^+, 5.5 mEq/L; Cl^-, 74 mEq/L; CO_2 content, 15 mmoles/L; and Δ, 39 mEq/L) represents the values about two weeks after the previous study. At this point the patient has a severe metabolic acidosis but the $[HCO_3^-]$ and pCO_2 are higher than expected, which leads us to conclude that some degree of metabolic alkalosis or respiratory acidosis is present. The former may be explained by the fact that the patient had been receiving HCO_3^- therapy for five days prior to the blood sampling; the latter may be explained by fatigue of the ventilatory muscles from the prolonged response to metabolic acidosis.

This patient illustrates an extremely complicated and unusual acid-base disorder. *The important point is that reliance upon the acid-base diagram of Figure 12-1 alone can lead one far astray if other factors, such as the clinical history,*

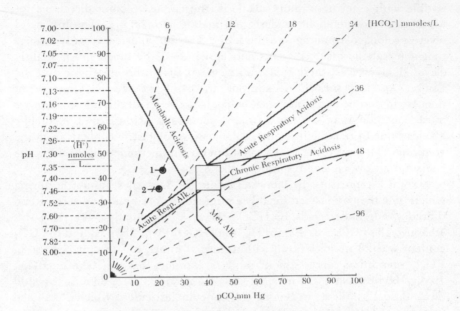

Figure 12-1D. Composite acid-base diagram for Case D.

medications, and other laboratory data which bear on the problem, are not also considered.

Case D, our last example, is that of a 51-year-old man (Fig. 12-1*D*) admitted to the hospital because of a gunshot wound of the left chest. The bullet was lodged in the spinal column at the level of the 12th thoracic vertebra. An operation was performed on the day of admission and the bullet removed. The patient did well for the first 48 hours after surgery but then proceeded to become drowsy. This change in consciousness progressed until he finally lost consciousness on the third postoperative day.

The blood chemistry data on the first postoperative day were as follows: [H^+], 42.5 nM; pCO_2, 21 mm Hg; [HCO_3^-], 12.2 mM; BUN, 28 mg/100 ml; CO_2 content, 14 mmoles/L; Cl^-, 106 mEq/L; Na^+, 141 mEq/L; K^+, 5 mEq/L; and Δ, 21 mEq/L. The elevated "anion gap" and slightly elevated BUN suggest a metabolic acidosis as does the low [HCO_3^-]. However, as shown by point 1 of Figure 12-1*D*, the pCO_2 is lower than would be expected for a pH of 7.37, and a respiratory alkalosis is probably also present.

The data on the third postoperative day were (point 2, Fig. 12-1*D*): [H^+], 36.3 nM; pCO_2, 20 mm Hg; [HCO_3^-], 13.2 mM; BUN, 66 mg/100 ml; creatinine, 2.9 mg/100 ml; CO_2 content, 13 mmoles/L; Cl^-, 111 mEq/L; Na^+, 149 mEq/L; K^+, 4.5 mEq/L, and Δ, 25 mEq/L. These data are even more suggestive of a metabolic acidosis due to renal failure combined with a respiratory alkalosis (causes of respiratory alkalosis are detailed in Chapter 11). In this case, the respiratory alkalosis was caused by a gram negative bacterial infection of the patient's surgical wound. Careful analysis of the acid-base data was the earliest sign that some other process was complicating the renal failure.

In summary then, analysis of the disturbances in acid-base balance requires an understanding of acid-base physiology and its variations in disease states. It also requires additional information such as other blood chemistry (e.g., electrolyte concentrations, BUN, and so forth), and clinical findings. Any of this information when considered as an isolated item can be misleading. However, when considered together these laboratory data and clinical findings can unravel a mysterious clinical picture.

COMMON LOGARITHMS

$$y = \log_{10} x$$

x	0	1	2	3	4	5	6	7	8	9	1	2	3	4	5	6	7	8	9
														Average differences					
10	0000	0043	0086	0128	0170	0212	0253	0294	0334	0374	4	8	12	17	21	25	29	33	37
11	0414	0453	0492	0531	0569	0607	0645	0682	0719	0755	4	8	11	15	19	23	26	30	34
12	0792	0828	0864	0899	0934	0969	1004	1038	1072	1106	3	7	10	14	17	21	24	28	31
13	1139	1173	1206	1239	1271	1303	1335	1367	1399	1430	3	6	10	13	16	19	23	26	29
14	1461	1492	1523	1553	1584	1614	1644	1673	1703	1732	3	6	9	12	15	18	21	24	27
15	1761	1790	1818	1847	1875	1903	1931	1959	1987	2014	3	6	8	11	14	17	20	22	25
16	2041	2068	2095	2122	2148	2175	2201	2227	2253	2279	3	5	8	11	13	16	18	21	24
17	2304	2330	2355	2380	2405	2430	2455	2480	2504	2529	2	5	7	10	12	15	17	20	22
18	2553	2577	2601	2625	2648	2672	2695	2718	2742	2765	2	5	7	9	12	14	16	19	21
19	2788	2810	2833	2856	2878	2900	2923	2945	2967	2989	2	4	7	9	11	13	16	18	20
20	3010	3032	3054	3075	3096	3118	3139	3160	3181	3201	2	4	6	8	11	13	15	17	19
21	3222	3243	3263	3284	3304	3324	3345	3365	3385	3404	2	4	6	8	10	12	14	16	18
22	3424	3444	3464	3483	3502	3522	3541	3560	3579	3598	2	4	6	8	10	12	14	15	17
23	3617	3636	3655	3674	3692	3711	3729	3747	3766	3784	2	4	6	7	9	11	13	15	17
24	3802	3820	3838	3856	3874	3892	3909	3927	3945	3962	2	4	5	7	9	11	12	14	16
25	3979	3997	4014	4031	4048	4065	4082	4099	4116	4133	2	3	5	7	9	10	12	14	15
26	4150	4166	4183	4200	4216	4232	4249	4265	4281	4298	2	3	5	7	8	10	11	13	15
27	4314	4330	4346	4362	4378	4393	4409	4425	4440	4456	2	3	5	6	8	9	11	13	14
28	4472	4487	4502	4518	4533	4548	4564	4579	4594	4609	2	3	5	6	8	9	11	12	14
29	4624	4639	4654	4669	4683	4698	4713	4728	4742	4757	1	3	4	6	7	9	10	12	13
30	4771	4786	4800	4814	4829	4843	4857	4871	4886	4900	1	3	4	6	7	9	10	11	13
31	4914	4928	4942	4955	4969	4983	4997	5011	5024	5038	1	3	4	6	7	8	10	11	12
32	5051	5065	5079	5092	5105	5119	5132	5145	5159	5172	1	3	4	5	7	8	9	11	12
33	5185	5198	5211	5224	5237	5250	5263	5276	5289	5302	1	3	4	5	6	8	9	10	12
34	5315	5328	5340	5353	5366	5378	5391	5403	5416	5428	1	3	4	5	6	8	9	10	11
35	5441	5453	5465	5478	5490	5502	5514	5527	5539	5551	1	2	4	5	6	7	9	10	11
36	5563	5575	5587	5599	5611	5623	5635	5647	5658	5670	1	2	4	5	6	7	8	10	11
37	5682	5694	5705	5717	5729	5740	5752	5763	5775	5786	1	2	3	5	6	7	8	9	10
38	5798	5809	5821	5832	5843	5855	5866	5877	5888	5899	1	2	3	5	6	7	8	9	10
39	5911	5922	5933	5944	5955	5966	5977	5988	5999	6010	1	2	3	4	5	7	8	9	10
40	6021	6031	6042	6053	6064	6075	6085	6096	6107	6117	1	2	3	4	5	6	8	9	10
41	6128	6138	6149	6160	6170	6180	6191	6201	6212	6222	1	2	3	4	5	6	7	8	9
42	6232	6243	6253	6263	6274	6284	6294	6304	6314	6325	1	2	3	4	5	6	7	8	9
43	6335	6345	6355	6365	6375	6385	6395	6405	6415	6425	1	2	3	4	5	6	7	8	9
44	6435	6444	6454	6464	6474	6484	6493	6503	6513	6522	1	2	3	4	5	6	7	8	9
45	6532	6542	6551	6561	6571	6580	6590	6599	6609	6618	1	2	3	4	5	6	7	8	9
46	6628	6637	6646	6656	6665	6675	6684	6693	6702	6712	1	2	3	4	5	6	7	7	8
47	6721	6730	6739	6749	6758	6767	6776	6785	6794	6803	1	2	3	4	5	5	6	7	8
48	6812	6821	6830	6839	6848	6857	6866	6875	6884	6893	1	2	3	4	5	5	6	7	8
49	6902	6911	6920	6928	6937	6946	6955	6964	6972	6981	1	2	3	4	4	5	6	7	8
50	6990	6998	7007	7016	7024	7033	7042	7050	7059	7067	1	2	3	3	4	5	6	7	8
51	7076	7084	7093	7101	7110	7118	7126	7135	7143	7152	1	2	3	3	4	5	6	7	8
52	7160	7168	7177	7185	7193	7202	7210	7218	7226	7235	1	2	2	3	4	5	6	7	7
53	7243	7251	7259	7267	7275	7284	7292	7300	7308	7316	1	2	2	3	4	5	6	6	7
54	7324	7332	7340	7348	7356	7364	7372	7380	7388	7396	1	2	2	3	4	5	6	6	7
x	0	1	2	3	4	5	6	7	8	9	1	2	3	4	5	6	7	8	9

Common logarithms continued

$$y = \log_{10} x$$

x	0	1	2	3	4	5	6	7	8	9	1	2	3	4	5	6	7	8	9
											\multicolumn{9}{c}{Average differences}								
55	7404	7412	7419	7427	7435	7443	7451	7459	7466	7474	1	2	2	3	4	5	5	6	7
56	7482	7490	7497	7505	7513	7520	7528	7536	7543	7551	1	2	2	3	4	5	5	6	7
57	7559	7566	7574	7582	7589	7597	7604	7612	7619	7627	1	2	2	3	4	5	5	6	7
58	7634	7642	7649	7657	7664	7672	7679	7686	7694	7701	1	1	2	3	4	4	5	6	7
59	7709	7716	7723	7731	7738	7745	7752	7760	7767	7774	1	1	2	3	4	4	5	6	7
60	7782	7789	7796	7803	7810	7818	7825	7832	7839	7846	1	1	2	3	4	4	5	6	6
61	7853	7860	7868	7875	7882	7889	7896	7903	7910	7917	1	1	2	3	4	4	5	6	6
62	7924	7931	7938	7945	7952	7959	7966	7973	7980	7987	1	1	2	3	3	4	5	6	6
63	7993	8000	8007	8014	8021	8028	8035	8041	8048	8055	1	1	2	3	3	4	5	5	6
64	8062	8069	8075	8082	8089	8096	8102	8109	8116	8122	1	1	2	3	3	4	5	5	6
65	8129	8136	8142	8149	8156	8162	8169	8176	8182	8189	1	1	2	3	3	4	5	5	6
66	8195	8202	8209	8215	8222	8228	8235	8241	8248	8254	1	1	2	3	3	4	5	5	6
67	8261	8267	8274	8280	8287	8293	8299	8306	8312	8319	1	1	2	3	3	4	5	5	6
68	8325	8331	8338	8344	8351	8357	8363	8370	8376	8382	1	1	2	3	3	4	4	5	6
69	8388	8395	8401	8407	8414	8420	8426	8432	8439	8445	1	1	2	2	3	4	4	5	6
70	8451	8457	8463	8470	8476	8482	8488	8494	8500	8506	1	1	2	2	3	4	4	5	6
71	8513	8519	8525	8531	8537	8543	8549	8555	8561	8567	1	1	2	2	3	4	4	5	5
72	8573	8579	8585	8591	8597	8603	8609	8615	8621	8627	1	1	2	2	3	4	4	5	5
73	8633	8639	8645	8651	8657	8663	8669	8675	8681	8686	1	1	2	2	3	4	4	5	5
74	8692	8698	8704	8710	8716	8722	8727	8733	8739	8745	1	1	2	2	3	4	4	5	5
75	8751	8756	8762	8768	8774	8779	8785	8791	8797	8802	1	1	2	2	3	3	4	5	5
76	8808	8814	8820	8825	8831	8837	8842	8848	8854	8859	1	1	2	2	3	3	4	5	5
77	8865	8871	8876	8882	8887	8893	8899	8904	8910	8915	1	1	2	2	3	3	4	4	5
78	8921	8927	8932	8938	8943	8949	8954	8960	8965	8971	1	1	2	2	3	3	4	4	5
79	8976	8982	8987	8993	8998	9004	9009	9015	9020	9025	1	1	2	2	3	3	4	4	5
80	9031	9036	9042	9047	9053	9058	9063	9069	9074	9079	1	1	2	2	3	3	4	4	5
81	9085	9090	9096	9101	9106	9112	9117	9122	9128	9133	1	1	2	2	3	3	4	4	5
82	9138	9143	9149	9154	9159	9165	9170	9175	9180	9186	1	1	2	2	3	3	4	4	5
83	9191	9196	9201	9206	9212	9217	9222	9227	9232	9238	1	1	2	2	3	3	4	4	5
84	9243	9248	9253	9258	9263	9269	9274	9279	9284	9289	1	1	2	2	3	3	4	4	5
85	9294	9299	9304	9309	9315	9320	9325	9330	9335	9340	1	1	2	2	3	3	4	4	5
86	9345	9350	9355	9360	9365	9370	9375	9380	9385	9390	1	1	2	2	3	3	4	4	5
87	9395	9400	9405	9410	9415	9420	9425	9430	9435	9440	0	1	1	2	2	3	3	4	4
88	9445	9450	9455	9460	9465	9469	9474	9479	9484	9489	0	1	1	2	2	3	3	4	4
89	9494	9499	9504	9509	9513	9518	9523	9528	9533	9538	0	1	1	2	2	3	3	4	4
90	9542	9547	9552	9557	9562	9566	9571	9576	9581	9586	0	1	1	2	2	3	3	4	4
91	9590	9595	9600	9605	9609	9614	9619	9624	9628	9633	0	1	1	2	2	3	3	4	4
92	9638	9643	9647	9652	9657	9661	9666	9671	9675	9680	0	1	1	2	2	3	3	4	4
93	9685	9689	9694	9699	9703	9708	9713	9717	9722	9727	0	1	1	2	2	3	3	4	4
94	9731	9736	9741	9745	9750	9754	9759	9763	9768	9773	0	1	1	2	2	3	3	4	4
95	9777	9782	9786	9791	9795	9800	9805	9809	9814	9818	0	1	1	2	2	3	3	4	4
96	9823	9827	9832	9836	9841	9845	9850	9854	9859	9863	0	1	1	2	2	3	3	4	4
97	9868	9872	9877	9881	9886	9890	9894	9899	9903	9908	0	1	1	2	2	3	3	4	4
98	9912	9917	9921	9926	9930	9934	9939	9943	9948	9952	0	1	1	2	2	3	3	4	4
99	9956	9961	9965	9969	9974	9978	9983	9987	9991	9996	0	1	1	2	2	3	3	3	4
x	0	1	2	3	4	5	6	7	8	9	1	2	3	4	5	6	7	8	9

2

ANSWERS TO PROBLEMS

CHAPTER 1

Problem 1: $[H^+] = 2 \times 10^{-5}$ M

$\log$ of $2 = 0.30$ $\log 10^{-5} = -5.0$

$\log 2 \times 10^{-5} = \log 2 + \log 10^{-5} = -4.7$

$pH = -\log H^+ = -(-4.7) = 4.7$

Problem 2: If the pH is 7.6, then $[H^+] = 10^{-7.6} = 10^{\overline{8}.4}$

$10^{\overline{8}.4} = 10^{-8} \times 10^{0.4} = 2.5 \times 10^{-8}$ M

2.5×10^{-8} M $= 25$ nmoles/L

Problem 3: 150 mEq H^+/L $= 1.5 \times 10^{-1}$ moles H^+/L

$\log$ of $1.5 = 0.17$

$\log 10^{-1} = -1$

$\log 1.5 \times 10^{-1} = \log 1.5 + \log 10^{-1} = -0.83$

$pH = -\log [H^+] = -(-0.83) = 0.83$

CHAPTER 2

Problem 1: $\quad \text{pH} = \text{pK}' + \log \dfrac{[\text{acetate}]}{[\text{acetic acid}]}$

$\quad\quad\quad\quad \text{pH} = \text{pK}' - \log \dfrac{[\text{acetic acid}]}{[\text{acetate}]}$

$\quad\quad\quad\quad 4.5 = 4.7 - \log \dfrac{[\text{acetic acid}]}{[\text{acetate}]}$

$\quad\quad\quad\quad \log \dfrac{[\text{acetic acid}]}{[\text{acetate}]} = 0.2$

$\quad\quad\quad\quad \dfrac{[\text{acetic acid}]}{[\text{acetate}]} = 1.6$

$\quad\quad\quad\quad \text{acetic acid (mmoles/L)} = \dfrac{1.6}{2.6} \times 100 \text{ mmoles/L}$

$\quad\quad\quad\quad\quad\quad\quad\quad\quad\quad\quad\; = 61 \text{ mmoles/L}$

$\quad\quad\quad\quad \text{acetate (mmoles/L)} = 100 - 61 = 39 \text{ mmoles/L}$

Problem 2: At pH 5, the mmoles of acetate/L would be the following:

$\quad\quad\quad\quad \text{pH} = \text{pK}' + \log \dfrac{[\text{acetate}]}{[\text{acetic acid}]}$

$\quad\quad\quad\quad 5.0 = 4.7 + \log \dfrac{[\text{acetate}]}{[\text{acetic acid}]}$

$\quad\quad\quad\quad \log \dfrac{[\text{acetate}]}{[\text{acetic acid}]} = 0.3$

$\quad\quad\quad\quad \dfrac{[\text{acetate}]}{[\text{acetic acid}]} = 2.0$

mmoles acetate/L $= 2/3 \times 100 = 67$ mmoles/L

Therefore to change the pH from 4.5 to 5.0
$67 - 39 = 28$ mmoles acetic acid is converted to acetate;
to do this 28 mmoles NaOH are required.
1 N NaOH contains 1 mmole NaOH/ml;
therefore 28 ml NaOH are needed.

Problem 3: $[H^+][OH^-] = k_w$

$k_w = 10^{-14}$ moles^2/L^2

At pH 5.0, $[H^+] = 10^{-5}$ moles/L

$[10^{-5}$ moles/L$][OH^-] = 10^{-14}$ moles^2/L^2

$\therefore [OH^-] = 10^{-9}$ moles/L

CHAPTER 3

Problem 1: $pH = pK' + \log \dfrac{[HPO_4^-]}{[H_2PO_4^-]}$

$7.4 = 6.8 + \log \dfrac{[HPO_4^-]}{[H_2PO_4^-]}$

$\log \dfrac{[HPO_4^-]}{[H_2PO_4^-]} = 0.6$

molar ratio $HPO_4^-/H_2PO_4^- = 3.98$ or ~ 4

Problem 2:   100 ml of 0.02 M phosphate buffer solution contains (100 ml) × (0.02 mmoles/ml) or 2 mmoles of phosphate buffer. At pH 7.4 the molar ratio $HPO_4^-/H_2PO_4^- = 4$ and therefore the 100 ml of buffer contains 0.4 mmoles of $H_2PO_4^-$ (1/5 × 2 mmoles).

At pH 6.8 the molar ratio $HPO_4^-/H_2PO_4^- = 1$ and therefore the 100 ml of buffer contains 1 mmole of $H_2PO_4^-$. Therefore to titrate the buffer from pH 7.4 to pH 6.8, 0.6 mmole of HPO_4^- is converted to $H_2PO_4^-$ and this requires 0.6 mmole of HCl. A 0.1 N HCl solution contains 0.1 mmole of HCl/ml; therefore 6 ml of 0.1 N HCl are required for this titration.

Problem 3:   .$pH = pK' + \log \dfrac{[A^-]}{[HA]}$

$7.4 = pK' + \log 1.2$

$pK' = 7.32$

CHAPTER 4

Problem 1: $pH = pK' + \log \dfrac{[HCO_3^-]}{0.03 \ pCO_2}$

$7.4 = 6.1 + \log [HCO_3^-] - \log 1.08$

$\log [HCO_3^-] = 1.33$

$[HCO_3^-] = 21.4 \ \text{mmoles/L}$

Problem 2: From the pCO_2 40 mm Hg isobar in Figure 4-6 it can be seen that when the pH falls to 7.3 from 7.4, the $[HCO_3^-]$ falls from 24 to 19 mmoles/L; therefore, 5 mmoles of H^+ from HCl must have been buffered by the bicarbonate buffer system. From the nonbicarbonate buffer curve *B–C* it can be seen that when the pH shifts from 7.4 to 7.3, Buf^- has accepted 2.5 mmoles of H^+ from HCl per liter of blood.

Therefore, for this pH change, 7.5 mmoles of HCl must have been added to the 1 liter of blood.

Problem 3: $pH = pK' + \log \dfrac{[HCO_3^-]}{0.03 \ pCO_2}$

$pH = 6.1 + \log \dfrac{20}{1.2}$

pH of the plasma $= 7.32$

$pH = pK' + \log \dfrac{[HPO_4^=]}{[H_2PO_4^-]}$

$7.32 = 6.8 + \log \dfrac{[HPO_4^=]}{[H_2PO_4^-]}$

$\log \dfrac{[HPO_4^-]}{[H_2PO_4^-]} = 0.52$

molar ratio $\dfrac{[HPO_4^=]}{[H_2PO_4^-]} = 3.3$

$[HPO_4^=] = 3.3/4.3 \times 2 = 1.53 \ \text{mmoles/L}$

$[H_2PO_4^-] = 1/4.3 \times 2 = 0.47 \ \text{mmoles/L}$

CHAPTER 5

Problem 1:   Assuming a barometric pressure of 760 mm Hg, then the alveolar air contains

$$\frac{40}{760} \times 100 = 5.26\% \; CO_2$$

Therefore 4000 ml/min $\times$ 0.0526 = 210 ml CO_2 are excreted per min. If his alveolar ventilation is reduced to 3 L/min and his metabolic rate remains the same, he will still excrete 210 ml CO_2/min once a new steady state is reached; under these conditions his per cent CO_2 in alveolar air is equal to

$$\frac{210 \text{ ml/min}}{3000 \text{ ml/min}} \times 100 = 7.0\%$$

If the barometric pressure remains 760 mm Hg then his alveolar pCO_2 becomes $0.07 \times 760 = 53.2$ mm Hg; his arterial blood pCO_2 will also be 53.2 mm Hg.

Problem 2:   Assuming (1) that within two minutes the alveolar pCO_2 will be about 53 mm Hg, (2) that at such time the blood has had little interaction with the interstitital fluid in regard to its new acid-base state, and (3) that the blood of this man is similar to that described in Figure 4-6, it can be estimated from Figure 4-6 that his arterial blood should have a pH of approximately 7.32. Check the accuracy of this estimation by using this pH and $[HCO_3^-]$ from the graph to calculate the pCO_2.

Problem 3:   The pCO_2 of the arterial blood will be 53.2 mm Hg. Assuming (1) that by 30 minutes blood and interstitial fluid have totally interacted in regard to the new acid-base state and (2) that the extracellular fluid buffering system of this man is similar to that described in Figure 4-7, it can be estimated from Figure 4-7 that his arterial plasma should have a pH of approximately 7.30. Check the accuracy of this estimation by using this pH and the $[HCO_3^-]$ from the graph to calculate the pCO_2.

CHAPTER 6

Problem 1:   At pH 4.4, the urine contains 40 μmoles of H^+ per liter. If he had to excrete 50 mEq of noncarbonic acid per day in the absence of a buffer system of NH_4^+ he would have to excrete 50,000 μmoles of free H^+ in the urine per day. That would take of daily urine volume of

$$\frac{50,000 \ \mu\text{moles } H^+}{40 \ \mu\text{moles } H^+/\text{L urine}} = 1250 \text{ L of urine/day}$$

Clearly impossible!

Problem 2: The kidney filtered 2.5 mEq HCO_3^-/min $\times$ 60 $\times$ 24 = 3600 mEq HCO_3^-/day. If the urine has a pCO_2 of 40 mm Hg and a pH of 5.1, it contains 0.12 mmoles of HCO_3^-/L, or the daily excretion of HCO_3^- in the urine is 0.24 mEq which is negligible compared to what is filtered, and thus it is fair to conclude that all HCO_3^- filtered is reabsorbed, i.e., 3600 mEq HCO_3^- is reabsorbed per day.

For every mEq of titratable acid or NH_4^+ in the urine the kidney has generated a HCO_3^- and delivered it to the peritubular blood; therefore, 250 mEq HCO_3^- was so generated per day. Thus each day the kidney delivers to the peritubular blood 3600 mEq HCO_3^- plus 250 mEq HCO_3^- or 3850 mEq HCO_3^- in this individual.

Problem 3: At this urinary pH, as stated in problem 2, the rate of urinary HCO_3^- excretion is so small it can be neglected.

At pH 7.4 the $HPO_4^=/H_2PO_4^-$ is

$$pH = pK' + \log \frac{[HPO_4^=]}{[H_2PO_4^-]}$$

$$7.4 = 6.8 + \log \frac{[HPO_4^=]}{[H_2PO_4^-]}$$

$$\log \frac{[HPO_4^=]}{H_2PO_4^-} = 0.6$$

$$\frac{HPO_4^=}{H_2PO_4^-} = 4$$

If it is assumed that the plasma pH is 7.4 then when the 50 mEq of phosphate buffer appearing in the urine was filtered, 40 mEq was in the form of $HPO_4^=$ and 10 mEq in the form of $H_2PO_4^-$.

However, in the urine with a pH of 5.1, it is in the following form:

$$pH = pK' - \log \frac{[H_2PO_4^-]}{[HPO_4^=]}$$

$$5.1 = 6.8 - \log \frac{[H_2PO_4^-]}{[HPO_4^=]}$$

$$\log \frac{[H_2PO_4^-]}{[HPO_4^=]} = 1.7$$

$$\frac{[H_2PO_4^-]}{[HPO_4^=]} = 50$$

Therefore approximately 49 mEq is present as $H_2PO_4^-$ and 1 mEq as $HPO_4^=$.

Therefore the tubular H^+ secretory system converted (49 mEq − 10 mEq) or 39 mEq of $HPO_4^=$ to $H_2PO_4^-$. It also converted 75 mmoles of NH_3 to NH_4^+. Therefore the tubular H^+ secretory system is eliminating H^+ from the body at the rate of 114 mmoles/day.

CHAPTER 7

Problem 1:   A net amount of alkali; 100 mEq per day

Problem 2:   Assuming his plasma pH to be 7.4:

$$pH = pK' + \log \frac{[HPO_4^=]}{[H_2PO_4^-]}$$

$$7.4 = 6.8 + \log \frac{[HPO_4^=]}{[H_2PO_4^-]}$$

$$\log \frac{[HPO_4^=]}{[H_2PO_4^-]} = 0.6$$

$$\frac{[HPO_4^=]}{[H_2PO_4^-]} = 4$$

Therefore, of the 50 mmoles of phosphate which appears in the urine, 40 mmoles were filtered as $HPO_4^=$ and 10 mmoles as $H_2PO_4^-$.

However, at a urine pH of 7.8:

$$7.8 = 6.8 + \log \frac{[HPO_4^=]}{[H_2PO_4^-]}$$

$$\log \frac{[HPO_4^=]}{[H_2PO_4^-]} = 1.0$$

$$\frac{[HPO_4^=]}{[H_2PO_4^-]} = 10$$

4.5 mmoles of phosphate are in the form of $H_2PO_4^-$ and 45.5 as $HPO_4^=$. Therefore, in forming the daily urine the kidney has titrated $(45.5 - 40)$ or 5.5 mmoles of $H_2PO_4^- \rightarrow HPO_4^=$ and in this way has excreted 5.5 mEq of alkali.

Problem 3: Since to be in a steady state he must excrete 100 mEq of alkali per day and only 5.5 mEq is being excreted by alkalinizing the phosphate buffer, he must be excreting 94.5 mEq of HCO_3^- per day.

INDEX

Acid(s), and bases, 13–17
  and buffers, chemistry of, 12–24
    in mammals, 25–39
    definitions of, 12–13
  and regulation of hydrogen ion concentrations, 8
  Brønsted-Lowry, 12, 13
  physiological, 25
  fixed, 26, 30, 31. See also *Noncarbonic acid.*
  Lewis concept of, 13
  noncarbonic. See *Noncarbonic acid.*
  strong and weak, and bases, 13–17
  titration curves of, 17–19
Acid-base regulation, and alveolar ventilation, 62
  blood and, 29
  definition of, 1
  principles of, clinical application of, 122–131
Acidemia, 84
Acidosis, 83
  diabetic, 84
  metabolic, 83–93
    acute and chronic, courses of, 87–93
    and respiratory alkalosis, 130
    buffering mechanisms of extracellular fluids in, 85–86
    buffering mechanisms of intracellular fluid in, 87
    causes of, 84–85
    definitions of, 83–84
    hyperventilation and, 115
    mixed, 84
      and metabolic alkalosis, 129
    renal mechanisms in, 87
    respiratory compensation in, 86
      limits of, 88
    simple, 84
    urinary pH and, 76, 87
  renal tubular, 85
  respiratory, 62, 104–113
    acute and chronic, 107–113
    and bicarbonate reabsorption, 81
    anion gap and, 123

Acidosis (*Continued*)
  respiratory, buffering mechanisms in, 105–106
    causes of, 104–105
    chronic, and metabolic alkalosis, 101, 127
    definition of, 104
    renal compensation in, 106–113
Aldosterone, and metabolic alkalosis, 97, 98
Alkalemia, 84
Alkali, and base, 12
  buffering of, 80
  elimination of, 80–82
  in metabolic alkalosis, 94
  processing of, by body, 79–82
  production of, sources of, 79–80
Alkalosis, 83
  "contraction," 95
  metabolic, 94–103
    and mixed metabolic acidosis, 129
    buffering mechanisms in, 97–98
    causes of, 94–97
    chronic respiratory acidosis and, 127
    definition of, 94
    renal response in, 99–103
    respiratory compensation in, 98–99
    signs and symptoms, 97
    treatment for, 100
  respiratory, 62, 114–121
    acute and chronic, courses of, 116–121
    and metabolic acidosis, 130
    and metabolic alkalosis, 103
    anion gap and, 123
    buffering in, 115–116
    causes of, 114–115
    definition of, 114
    renal response to, 116
Alveolar hyperventilation, 38, 114, 115
Alveolar hypoventilation, 104, 105
Alveolar ventilation
  and acid-base regulation, 62
  and elimination of carbon dioxide, 61
  and partial pressure carbon dioxide level, 63
  definition, 62

143